NONLOCALITY IN QUANTUM PHYSICS

NONLOCALITY IN QUANTUM PHYSICS

ANDREY ANATOLJEVICH GRIB
State University of Economics
and Finances of St. Petersburg
St. Petersburg, Russia

WALDYR ALVES RODRIGUES, JR.
State University of Campinas
and Salesian University
Campinas, Brazil

Kluwer Academic/Plenum Publishers
New York, Boston, Dordrecht, London, Moscow

Library of Congress Cataloging-in-Publication Data

Grib, Andreĭ Anatol'evich.
  Nonlocality in quantum physics / Andrey Anatoljevich Grib and
Waldyr Alves Rodrigues, Jr.
     p.   cm.
  Includes bibliographical references and index.
  ISBN 0-306-46182-X
  1. Quantum theory.   2. Physics--Philosophy.   I. Rodrigues, W. A.
II. Title.
QC174.2.G7468 1999
530.12--dc21   99-33174
                                                               CIP

ISBN: 0-306-46182-X

© 1999 Kluwer Academic / Plenum Publishers
233 Spring Street, New York, N.Y. 10013

A C.I.P. record for this book is available from the Library of Congress.

10 9 8 7 6 5 4 3 2 1

All rights reserved

No part of this book may be reproduced, stored in a retrieval system, or transmitted in any form or by any means, electronic, mechanical, photocopying, microfilming, recording, or otherwise, without written permission from the Publisher

Printed in the United States of America

Preface

The nonlocality phenomena exhibited by entangled quantum systems are certainly one of the most extraordinary aspects of quantum theory. This book discusses this phenomenon according to several points of view, i.e., according to different interpretations of the mathematics of the quantum formalism.

The several interpretations of the Copenhagen interpretation, the many worlds, the de Broglie-Bohm, quantum logics, the decohering by the environment approach and the histories approach interpretations are scrutinized and criticized in detail. Recent results on cryptography, quantum bit commitment, quantum erasers and teleportation are also presented and discussed.

In preparing the book we benefited from discussions with many people, but we would like, in particular, to express our gratitude to Professor B. d'Espagnat for his useful comments and suggestions. We are grateful also to Ms. L. Gentry El-Dash for the English revision, to Dr. J. E. Maiorino for the production of the figures and a careful reading of the manuscript, and for the staff of Plenum for advice and for having produced a nice book. Finally, the authors thank FAPESP (contract no.1996/2657-0) for a grant making this book possible.

A. A. GRIB AND W. A. RODRIGUES, JR.

Contents

1.	INTRODUCTION	1
	Chapter References	5
2.	FUNDAMENTALS	7
	2.1 The rules of the game	7
	2.2 The whole and its parts	12
	2.3 Classical limit and decoherence	14
	2.3.1 Ehrenfest theorem	15
	2.3.2 Classical dynamical variables and collective observables	15
	2.3.3 The WKB approximation	19
	2.3.4 Decoherence and Schrödinger's cat paradox	20
	2.4 Proofs of the frequency interpretation	23
	Chapter References	26
3.	HOW Ψ COLLAPSES	29
	3.1 The problem	29
	3.2 Collapse and unitary evolution	30
	Chapter References	33
4.	BELL'S INEQUALITIES	35
	4.1 The Einstein, Podolski and Rosen (EPR) argument	35
	4.2 D'Espagnat's version of Bell's inequalities	36
	4.3 Stapp's formulation of Bell's inequalities	37
	4.4 Bell's proof	39
	Chapter References	41
5.	COPENHAGEN INTERPRETATION	43
	5.1 Absence of quantum objects	45
	5.1.1 Nonlocality	46
	5.1.2 Criticisms	47
	5.2 Objectively existing potentialities world	49
	5.2.1 Nonlocality	51

	5.2.2	Criticisms		53
		5.2.2.1	The Wigner friend paradox	53
		5.2.2.2	The problem of quantum processes before the existence of conscious observers	53
	Chapter References			55
6.	MANY WORLDS INTERPRETATION (MWI)			59
	6.1	The role of the observer		59
	6.2	Nonlocality		61
	6.3	Criticisms		62
	Chapter References			64
7.	QUANTUM LOGIC (QL)			67
	7.1	Proposal		67
		7.1.1	Weak modularity postulate and proposition calculus	70
		7.1.2	The Boolean lattice of a classical system	71
		7.1.3	The non distributive lattice associated to a quantum system	72
	7.2	Hasse diagrams		74
		7.2.1	Spin one system for which two spin projections, S_x and S_z, are measured	75
	7.3	The birth of time		76
	Chapter References			80
8.	MACROSCOPIC REALIZATIONS OF QL			83
	8.1	Lattices and graphs		83
		8.1.1	Grib and Zapatrin rule for construction of the lattice corresponding to a given graph and vice-versa	
	8.2	Breakdown of Bell's inequalities		91
	8.3	The Problem of Nonlocality		95
	Chapter References			95
9.	HIDDEN VARIABLES THEORIES (HVT)			97
	9.1	What is a hidden variables theory?		97
		9.1.1	Comments on HVT of the first kind	99
		9.1.2	Theories of the second and zeroth kinds	101
	9.2	Von Neumann's theorem		101
	9.3	Spin 1/2 hidden variable's model		103
	9.4	Gleason's theorem. Kochen-Specker paradox		104
		9.4.1	Gleason's theorem	105
		9.4.2	Kochen-Specker theorem	108
		9.4.3	Contextuality in HVT and the Mermin example	108
		9.4.4	Kochen-Specker paradox	110
	Chapter References			112
10.	DE BROGLIE-BOHM NON-RELATIVISTIC HVT			115
	10.1	Non relativistic de Broglie-Bohm theory		115
		10.1.1 Stationary states		117

    10.1.2 Non conservation of energy and momentum in non stationary states   117
  10.2 The many particle system   118
  10.3 Remarks on the non relativistic BBT   119
  Chapter References   121

11. DE BROGLIE-BOHM RELATIVISTIC HVT   123
  11.1 HVT interpretation of the boson equations   123
  11.2 The BBT interpretation of the Dirac equation   125
  11.3 Relativistic BBT many-fermion theory   126
  11.4 Necessity of superluminal processes   127
  Chapter References   129

12. STATISTICAL INTERPRETATION   133
  12.1 Introduction   133
  12.2 When is the SIQT possible?   135
  12.3 SIQT × orthodox quantum theory   137
  Chapter References   141

13. NON-UNITARY EVOLUTION?   143
  13.1 Prigogine and Ghirardi et al. theories   144
  13.2 Criticisms   146
  Chapter References   148

14. HISTORIES APPROACH   151
  14.1 What is a history?   151
  14.2 Consistent histories   152
  14.3 Criticisms   155
  Chapter References   157

15. QUANTUM MIRACLES AND CRYPTOGRAPHY   159
  15.1 Greenberger et al. example   159
  15.2 Hardy's example   161
  15.3 Negative experiments. Elitzur-Vaidman proposal   163
  15.4 Delayed choice experiments   164
  15.5 Quantum erasers   165
  15.6 Quantum cryptography   168
    15.6.1 Quantum bit commitment   169
  15.7 Teleportation   173
  Chapter References   178

16. WHERE ARE WE NOW?   181
  Chapter References   184

Appendices
A– SET THEORY AND LATTICES   187

A.1	Main definitions	187
A.2	Posets	189
A.3	Lattices	190
A.4	The Boolean algebra of sets	194
A.5	Borel sets, measures and integration	194
	Chapter References	199

B– HILBERT SPACES — 201

B.1	Definition of Hilbert Space	201
B.2	Linear manifolds and subspaces	203
	B.2.1 The lattice of subspaces	207
B.3	Dual and conjugate dual spaces	207
B.4	Tensor products	209
B.5	Linear operators	210
B.6	Projection operators and lattice structure	213
	Chapter References	216

Index — 217

1
INTRODUCTION

We are approaching the end of the XX-th century and these extraordinary days witness an increasing interest among the members of the scientific community (and also among cultured laymen) in the foundations of quantum physics. This interest is mainly due to three different reasons.

(i) The recent developments in quantum optics and other experimental techniques in quantum physics that led to the possibility of exciting experiments, capable of testing the very foundations of the theory and even the philosophical principles underlying it.[1] Among these experiments are the ones proving the breakdown of Bell's inequalities, the delayed choice experiment, the quantum Zeno effect, some macroscopic quantum effects using Josephson junctions, quantum cryptography, teleportation, etc.

(ii) Many new fellows among the new generation of physicists are asking about *the meaning* of the formulas of quantum theory which lead to so many magnificent predictions of experimental facts not only in atomic, molecular and nuclear physics, but also in elementary particles physics[2]. This has the important consequence that the interpretations of quantum theory given by its "fathers"— N. Bohr, M. Born, W. Heisenberg, V. Fock, J. von Neumann, on one side and A. Einstein, E. Schrödinger, L. de Broglie and D. Bohm on the other side are now being discussed more intensely than ever before.

(iii) Success of the Big Bang theory in cosmology leads to the necessity of speaking of the earlier universe totally in terms of quantum concepts, because there were no macroscopic bodies before the so called *recombination era* and there is an opinion that the Big Bang itself, as originator of classical spacetime, must be treated as a quantum phenomenon. Also, the application of quantum theory to cosmology emphasizes the problem of the role of the observer in quantum physics.

The objective of our research is to discuss open-mindedly the different interpretations of quantum physics concerning the so-called problem of the *nonlocality* in quantum theory. This nonlocality is mainly connected with Einstein-Podolsky-Rosen (*EPR*) situations and is different from nonlocality as it is understood in the quantum field theory of elementary particles, where it means the breakdown of commutativity for local observables for spacelike intervals. Interactions of elementary particles are here supposed to be local and possible extensions to non local theories (e.g., for superstring theory, etc.) will not be discussed here. Also, with regard to quantum cosmology (and, in general, also in quantum field theory in curved spacetime) some new types of nonlocality (e.g., nonlocality in time for *time machines*) arise. These important and controversial issues are not discussed in this book.

As to the structure of the book we want to say the following. Our goal was to write a book as short as possible, because thick books usually are not read. Nevertheless, it is not a popular book; sometimes, when discussing topics, as, e.g., quantum logic, rather sophisticated mathematics is used! However, our interest was to discuss fundamental concepts, not calculation methods. And mathematics is used for clarification of these concepts, especially in cases where ordinary language is inadequate for expressing them. We hope that with the aid of the mathematical appendices the book can be read as some self contained subject without using additional literature, so that undergraduate and graduate students of physical and mathematical departments of universities can understand it.

In Chapter 2 we give a short presentation of the mathematical principles of the theory (*the rules of the game*) and introduce one of the more disconcerting issues of quantum theory, namely the *whole and its parts*. We introduce also in Chapter 2 the fundamental and yet open problem of the classical limit and the phenomenon of decoherence.

Chapter 3 is dedicated to the crucial problem associated with the wave packet collapse, which is discussed under several points of view in the next chapters. Chapter 4 is dedicated to a study of the different forms of Bell's inequalities and the meaning of their breakdown, which implies a well defined kind of nonlocality. Then, in Chapter 5 the *Copenhagen interpretation* of quantum physics is given. Much attention is paid to the not widely known modern versions of *nonobjective physics,* where either microobjects do not exist (only macroobjects exist, microobjects being the name for special relations between them), or there are no *micro* and *macroobjects* at all and only *acts* (*sharp* or *fuzzy*) are fundamental! These views developed by G. Ludwig, A. Bohr and D. Finkelstein (discussed in Chapter 5) are based on some interpretations of N. Bohr's words about the fundamental role of ordinary language, classical physics, classical apparatuses and our acts with them, when interpreting quantum physics. This is not a book on philosophy of physics, but surely different interpretations of quantum theory lead to different philosophical views and we sometimes make short comments about them.

Another *interpretation* of Copenhagen interpretation, as we claim here, is given by the Heisenberg-Fock *objectively existing potentialities world* so that in some sense quantum objects can be considered as *existing*! Close to it are the *many worlds interpretation* (Chapter 6) and the *quantum logic interpretation* (Chapter 7). Differently from many other books on the same topic, we pay much attention to the problem of the role of consciousness of the observer in the Copenhagen interpretation, following J. von Neumann, E. P. Wigner and F. London et al. Consciousness is used in our text as synonym for the subject that chooses, experiencing what it chooses, as it collapses the quantum state vector of a given quantum system. How this is done nobody knows yet, but our claim is that the role of consciousness is not to be attributed to some *mystical* (in the bad sense of the word *mystical* as something unclear and ambiguous) influence of it on the Universe, but must be mathematically expressible as resulting from the difference between the logic of the human *consciousness* which (at least in the awake state) is a Boolean one and the non-Boolean logic of the (quantum) world. In this way there is necessity of *translation* from one logic to the other when truth values are given by the observer, and this is manifested in the formalism of quantum theory. Nevertheless, the serious philosophical, probably unsolvable, problem, posed with the great force by the German philosopher I. Kant arises here. What in our vision of the Universe is conditioned by *a priori* forms of our reasoning and what belongs to the world itself?

Criticisms of the different versions of Copenhagen interpretations presented in Chapters 4, 5, 6 and 7 based on Schrödinger's cat paradox, Wigner's friend, preferable basis for many universes etc, are also discussed in the appropriate places.

In this part of the book (chapters 4, 5, 6, 7) much attention also is given to properties of complementarity expressed either in the noncommutativity of quantum observables or in the nondistributivity of the quantum logic lattice. But it is not only these typically quantum features that are important for understanding the physical meaning of the breakdown of Bell's inequalities and quantum nonlocality. Indeed, we show that a crucial role is played also by the principles of special relativity. That is the reason why in Chapter 8 we discuss some examples of *macroscopic* realizations of complementarity and nondistributive structures for which Bell's inequalities can also be broken, but without a manifestation of a nonlocality effect for spacelike intervals as occurs in microphysics.

These examples show the possibility of *imitating* quantum systems by some macroscopic automata, called by us *macroscopic quantum computers* in view of the possible application of them for quantum computations. Nevertheless, as shown in Chapter 8, for these systems, which are based on classical physical elements, quantum nonlocality cannot manifest itself, thus showing the real difference between quantum microobjects and a classical imitation of them obtained through the correspondence between the quantum logic lattice of the former and the graph of the latter.

All this leads us to discuss in Chapters 9, 10, 11 and 12 different versions of *hidden variables theories,* which consider quantum physics as a manifestation of some other (usually having such classical features as trajectories, etc.) fundamental physics different from it. After giving some basic theorems, like von Neumann's, Gleason's and Kochen-Specker's, we study with care the de Broglie-Bohm nonrelativistic and relativistic hidden variables theories. Some differences between these theories (relaxation time effect, etc.) and standard quantum mechanics which can be experimentally checked, are also discussed.

Certainly, the main difference between hidden variables theories and standard quantum theory arises in the relativistic realizations of the former. There, *differently* from what is often used in elementary particles physics, it is impossible to speak about boson particles. Only classical boson fields can exist, while fermions can be considered as particles. Another problem of hidden variables theories, which we discuss, is the one concerning the breakdown of Lorentz invariance due to the necessity of *superluminal processes* which makes necessary some changes in special relativity. In standard quantum theory we have the so called *passive nonlocality*, manifested in coincidences between random events conditioned by quantum indeterminism, while in hidden variables theory we have *active nonlocality*, breaking the usual relativistic causality at some level.

The *statistical interpretation*, considered as a hidden variables theory claiming that a given wave function is the description not of an individual quantum system but of an *ensemble* of identical quantum systems is shortly discussed in Chapter 12. The necessity of *changing* some rules of the standard probability theory is stressed.

In Chapter 13 we investigate and criticize the proposals of I. Prigogine and G. C. Ghirardi and collaborators to *change* Schrödinger's equation in such a way that a nonunitary evolution becomes possible.

Chapter 14 is dedicated to a short but critical appraisal of the *histories approach* which exposes both the hopes of their proponents as well as some unsustainable claims used in their approaches.

In Chapter 15 examples of quantum nonlocality different from those in the famous Aspect's experiment are investigated. Some non trivial proposals by D. M. Greenberger et al, L. Hardy, A. C. Elitzur and L. Vaidman and also the delayed choice *quantum miracles* are discussed. The main ideas concerning *teleportation* and *quantum cryptography* are also briefly discussed and it is made clear that when applied to the problem known as *bit commitment*, quantum cryptography is not safe, contrary to what was thought until recently.

In Chapter 16 we discuss *where we are now*, presenting our conclusions.

References are given at the end of each chapter. The authors must apologize for being too succinct either in the exposition of some of the interpretations of quantum theory[3] or in the presentation of references that clearly denounces some of their tastes, which of course are not always the same.

In spite of the fact that both authors are from different countries, Russia and Brazil, and belong to different schools of theoretical physics, they want to say here that they experienced a great joy while discussing in Campinas, Brazil during the very hot summer of 1997, the foundations of quantum physics. If something of this joy of participating in one of the greatest adventures of the twentieth century will pass to the readers, the authors will consider their goal achieved!

Notes

1. The philosophy behind quantum mechanics is discussed in many text books, such as, in[1–6]. As for the case of the physical interpretation, there is no general agreement among the authors on this issue.

2. Obviously there are still many problems concerning the formulation of a fundamental theory for elementary particles. Nevertheless, we think with the majority view that the principles of quantum physics must be at the foundations of the correct theory.

3. Of course, no claim is done that we have discussed all interpretations that exist in the literature.

References

[1] M. Jammer, *The Philosophy of Quantum Mechanics* (J. Wiley & Sons, New York, 1974).

[2] B. d'Espagnat, *Conceptual Foundations of Quantum Mechanics*, second edition (W. A. Benjamin Inc., Reading, MA, 1976).

[3] B. d'Espagnat, *Veiled Reality, An Analysis of Present-Day Quantum Concepts* (Addison-Wesley Publ. Co., Reading, MA, 1995).

[4] M. Redhead, *Incompleteness, Nonlocality, and Realism. A Prolegomenon to the Philosophy of Quantum Mechanics* (Clarendon Press, Oxford, 1987).

[5] F. Selleri, *Quantum Paradoxes and Physical Reality* (Kluwer Academic Publishers, Dordrecht, 1990).

[6] H. P. Stapp, *Mind, Matter and Quantum Mechanics* (Springer-Verlag, Berlin, 1994).

2
FUNDAMENTALS

In this Chapter we briefly recall the fundamentals of quantum physics that we shall need in what follows. In section 2.1 we give the rules of the game, in section 2.2 we discuss one of the more disconcerting issues of quantum theory, namely, the whole and its parts, and in section 2.3 we discuss the fundamental (yet open) problem of the classical limit of the theory and the phenomenon of decoherence. In section 2.4 we study some proposed proofs of the frequency interpretation.

2.1 THE RULES OF THE GAME

1. The *states* of quantum systems are mathematically described by *vectors*[1] in a Hilbert space, which will be denoted by $\mathcal{H}$ in what follows. This space is usually taken as complex, but other formulations are possible, e.g., we can have a real Hilbert space or quaternionic spaces [1-7]. More recent references are [8,9].

The *principle of superposition* (*PS*) is valid. This means that if the vectors $|\psi_1\rangle, |\psi_2\rangle, ..., |\psi_n\rangle \in \mathcal{H}$ are possible states of a given physical system, then any normalized vector $\Psi \in \mathcal{H}$,

$$|\Psi\rangle = c_1|\psi_1\rangle + c_2|\psi_2\rangle + ... + c_n|\psi_n\rangle \tag{2.1}$$

where $c_1, c_2, ..., c_n \in \mathcal{C}$ (the field of complex numbers) is also a possible state of the system[2].

2. To each observable physical quantity A of a given physical system there corresponds a Hermitian (self conjugate) operator $\mathbf{A} \in \mathcal{H} \otimes \mathcal{H}^\dagger = \mathcal{V}$ where $\mathcal{H}^\dagger$ is the so called *conjugate dual space*[10][3] of $\mathcal{H}$. The set of all operators corresponding to

7

the observables of a given physical system possesses a Lie algebra structure (so-called *observable algebra*), where the internal composition law is the commutator product. We say that two observables A and B are *compatible* if and only if the corresponding operators $\mathbf{A}$ and $\mathbf{B}$ commute, i.e., $[\mathbf{A}, \mathbf{B}] = 0$.

3. The outcome of a measurement of an observable A is always one of the eigenvalues of the operator $\mathbf{A}$.

4. The time evolution of the state vector $|\Psi(t)\rangle$ which describes the state of a given quantum system at time t is causal (and linear) if the system is not subjected to any measurement by some observer. More precisely, this means that from $|\Psi(t_0)\rangle = \sum_{i=1}^{n} c_i |\psi_i(t_0)\rangle$ we can calculate

$$|\Psi(t)\rangle = \sum_{i=1}^{n} c_i |\psi_i(t)\rangle \tag{2.2}$$

The evolution from $t_0 \mapsto t$ is given by the linear *Schrödinger equation*,

$$\mathbf{H}|\Psi(t)\rangle = i\hbar \frac{\partial |\Psi(t)\rangle}{\partial t} \tag{2.3}$$

where $\mathbf{H}$ is an Hermitian operator called the *Hamiltonian*. When it does not depend on time we have

$$|\Psi(t)\rangle = \exp\left[\frac{-i}{\hbar}\mathbf{H}(t - t_0)\right] |\Psi(t_0)\rangle \tag{2.4}$$

The norm of the state vector $|\Psi(t)\rangle$ is of course conserved for all time t because the *evolution operator*

$$U(t, t_0) = \exp\left[\frac{-i}{\hbar}\mathbf{H}(t - t_0)\right] \tag{2.5}$$

is unitary.

5. Postulate of *wave function collapse (or reduction of the state vector)* and *quantum indeterminism*.

When we measure a given observable A, a system with state vector $|\Psi(t)\rangle$ jumps indeterministically into one of the eingenvectors of $\mathbf{A}$. If

$$\mathbf{A}|u_n\rangle = a_n|u_n\rangle \tag{2.6}$$

then,

$$|\Psi\rangle \mapsto |u_n\rangle \tag{2.7}$$

The probability for $|\Psi\rangle = \sum c_m |u_m\rangle \mapsto |u_n\rangle$ is given by

$$W_n = |\langle u_n|\Psi\rangle|^2 = |c_n|^2 \tag{2.8}$$

The measurement of a given observable A in system S at $t = 0$ is said to be the *first kind or ideal*, also called von Neumann measurement, if it happens that if the measurement is repeated a second time it will give the same value as the first time. We can have two cases:

(i) If S at the time of the first measurement is in one of the eigenstates of $\mathbf{A}$, we can consider that the value of the quantity A obtained is the one it has *before* and *after* $t = 0$.

(ii) If S at the time of the first measurement is not in one of the eigenstates of $\mathbf{A}$, then we can consider the result of the first measurement as being the value of A after $t = 0$.

A measurement of a given observable A in a system S at $t = 0$ is said to be the *second kind or non ideal*, also called Landau measurement, if a second measurement will not give the same result as the previous measurement. The canonical example of this kind is given, e.g., by the measurement of the x component p_x of the momentum of a particle with an uncertainty $\triangle p_x$. The measuring apparatus in general modifies the value of the measured quantity, independently of the original state of the particle. It is obvious that after the measurement the particle (or an ensemble of particles) cannot be described by state vectors that are eigenvectors of the operator $\mathbf{p}_x$, with eigenvalues in the range $p_x \pm \triangle p_x$. This implies that a second measurement in general will give a result different from the first one. For *second kind* measurements we must distinguish two cases:

(i) If S is originally in an eigenstate of $\mathbf{A}$ we can consider the result of the measurement as the value of A before $t = 0$.

(ii) If S is not originally in an eigenstate of $\mathbf{A}$, then we cannot give any interpretation to the result of an individual measurement in terms of a property possessed by S.

Now, some remarks are necessary. If the observer did not register the position of the pointer of his apparatus to which corresponds one of the numbers a_n, but the measurement took place, he could say that W_n is the probability of a defined position of the pointer. This observer with *incomplete information* will describe the quantum system not by the vector $u_n \in \mathcal{H}$ called a *pure state* but by a *density matrix* $\rho \in \mathcal{H} \otimes \mathcal{H}^\dagger$. In this case the quantum system is said to be a *mixture of states* (or mixture, for simplicity). The density matrix, also called *statistical operator* is sometimes denoted by $\mathbf{P}_\rho$

$$\rho \equiv \mathbf{P}_\rho = \sum_n W_n \mathbf{P}_{u_n} \tag{2.9}$$

where $\mathbf{P}_{u_n} = \mathbf{P}_{u_n}^2$ are the *projectors (idempotent and self adjoint operators)* on the vectors $|u_n\rangle$. The matrix ρ has the following properties:
  (i) It is Hermitian, i.e., $\rho^\dagger = \rho$
  (ii) It is normalized, i.e., $Tr\rho = 1$
  (iii) It is positive definite, i.e., $\langle \Psi | \rho | \Psi \rangle \geq 0, \forall | \Psi \rangle \in \mathcal{H}$.

In the case of a given measurement made by an apparatus but not registered by the observer, the density matrix ρ is such that its eigenvalues are the probabilities W_n. So, in a mixture of states we may have the system in state $|u_i\rangle$, with probability W_i.

It is very important to have in mind that we also use density matrices for the description of *subsystems* of a given quantum system (see section 2.2), but in this case, and contrary to wide spread misunderstandings, they do not have the meaning of a mixture of states. On this point we must agree with d'Espagnat[11] who proposed to call these states *improper mixtures* in order to distinguish them from the ones where the density matrices really describe *proper* mixtures. To a vector $|\Psi\rangle \in \mathcal{H}$ describing, of course, a pure state of a given quantum system, there corresponds a unique special density matrix $\rho = \mathbf{P}_\Psi = |\Psi\rangle\langle\Psi|$ and obviously we have $\rho = \rho^2$

For this special case it is evident that

$$Tr\rho^2 = Tr\rho = 1, \qquad (2.10)$$

but in the general case of a mixture we always have

$$Tr\rho^2 \leq 1. \qquad (2.11)$$

The *proof* is as follows. Take $\mathbf{P}_\rho = \sum_n W_n \mathbf{P}_{u_n}$, with $0 \leq W_n \leq 1, \sum_n W_n = 1$. Suppose by absurd that we have a mixture and $Tr\rho^2 = 1$. This necessarily implies that $\sum_n W_n^2 = 1$ and since $\sum_n W_n = 1$ we get

$$\sum_n W_n(1 - W_n) = 0. \qquad (2.12)$$

Equation (2.12) implies that all but one of the W_n are equal to zero. The non-null value of course must be 1, i.e., we must have a pure state.

In many physical situations the density matrix has degenerate eigenvalues and in such cases not one, but many mixtures are described by it. An instructive example is the following. Consider a spin 1/2 particle and choose one of the orthonormal sets $\{|x^i, l\rangle\}, i = 1, 2, 3 ; l = 1, 2$, which are respectively eigenvectors of the spin operators $\sigma_1, \sigma_2, \sigma_3$ ($\sigma_i\sigma_j + \sigma_j\sigma_i = 2\delta_{ij}$), as a base for the spin space[4]. We have,

$$\sigma^i|x^i, l\rangle = (-1)^{l+1}|x^i, l\rangle. \qquad (2.13)$$

The algebraic properties of the spin operators entail the following relations between the different orthonormal basis,

$$|x^1, l\rangle = \frac{1}{\sqrt{2}}[|x^3, 1\rangle + (-1)^{l+1}|x^3, 2\rangle]; \\ |x^2, l\rangle = \frac{1}{\sqrt{2}}[|x^3, 1\rangle + i(-1)^{l+1}|x^3, 2\rangle]. \qquad (2.14)$$

The density matrix for an ensemble of *totally non-polarized* spin 1/2 particles can be determined at once. Indeed, if we perform an experiment to measure, e.g., the S_z component of the spin, then by definition to be totally unpolarized means that the two possible values will show the same frequency. We have, then

$$\rho = \frac{1}{2}(|x^3, 1\rangle\langle x^3, 1| + |x^3, 2\rangle\langle x^3, 2|). \qquad (2.15)$$

or, taking into account Eq. (2.14) we obtain,

$$\rho = \frac{1}{2}(|x^1, 1\rangle\langle x^1, 1| + |x^1, 2\rangle\langle x^1, 2|);$$

$$\rho = \frac{1}{2}(|x^2, 1\rangle\langle x^2, 1| + |x^2, 2\rangle\langle x^2, 2|);$$

$$\rho = 1/6(\sum_{i,l} |x^i, l\rangle\langle x^i, l|). \tag{2.16}$$

The matrix representation of ρ in the basis $\{|x^3, l\rangle\}, l = 1, 2$ is

$$\rho_{x^3} = \frac{1}{2}\begin{bmatrix} 1 & 0 \\ 0 & 1 \end{bmatrix} \tag{2.17}$$

These equations show that ρ in our example describes indeed different mixtures.

Postulate **5** of the wave function collapse is sometimes called *transformation of the pure state into a mixture in the measurement*. In reality, the registration by the observer of the position of a pointer is as in classical physics, i.e., we pass from incomplete to complete information. Now, postulate **5** presupposes some operation which according to von Neumann[13] is the following. Write the state vector $|\Psi\rangle$, describing an *ensemble* (see **6** below) of identically prepared systems, in some orthonormal basis of eigenvectors of the operator **A**,

$$|\Psi\rangle = \sum_n c_n |u_n\rangle. \tag{2.18}$$

To $|\Psi\rangle$ there is associated the density matrix $\rho = \mathbf{P}_\Psi$. Postulate **5** means that if we realize measurements of A on the whole ensemble and do not take care in separating the subensembles corresponding to the different eigenvalues, it will be described by a new density matrix resulting from the transformation

$$\rho = \mathbf{P}_\Psi \mapsto \rho' = \sum_n W_n P_{u_n}, \tag{2.19}$$

where $\rho^2 = \rho$, but $\rho'^2 \neq \rho', W_n = |c_n|^2$.

The transformation given by Eq. (2.19) cannot be described by the unitary evolution operator (Eq. 2.5), i.e., we cannot write $\rho' = \mathbf{U}\rho\mathbf{U}^{-1}$, which means that this transformation contradicts the Schrödinger equation and the linearity of the evolution.

So, in quantum physics there are two totally different ways of *changement* in time of the state vector describing a given physical system:

(i) *deterministic evolution* described by the Schrödinger equation (2.1);

(ii) *indeterministic*, by *jump* due to the wave function collapse (Eq. 2.19).

6. *The frequency interpretation.* If we do the same experiment (same experimental arrangement) many times, we can say that each individual experiment constitutes a *preparation* of a definite state and the set of identical experiments define an *ensemble*.

An example of an ensemble of quantum particles with the same state vector $|\Phi\rangle$ is given by an electronic pump which fires one electron after the other separated by some interval of time.

The frequency interpretation of the wave function (*Born Postulate*) consists in the following statement. If for the particles in the ensemble we measure the observable A represented by the operator $\mathbf{A}$ with eigenvectors $|u_n\rangle$ and eigenvalues a_n, then the frequency of the definite result a_n approaches the value

$$W_n = |\langle \Psi | u_n \rangle|^2, \tag{2.20}$$

if the number of elements in the ensemble is infinite.

With this postulate quantum theory makes clear the possibility of predicting for a quantum system prepared in the state $|\Psi\rangle$ (using a preparation device) the frequency of this or that result for some observable property A (through a measuring device) given an ensemble with an infinite number of quantum systems prepared in the same way.

This postulate looks much like von Mises' statistical definition of probability. Now, in classical physics we have the classical definition of probability (in terms of equal possible outcomes of the experiment) and the axiomatic (Kolmogorov's definition in terms of a probabilistic measure in the space of events (see, e.g., [14]).

In quantum theory the last two possibilities do not apply. Some authors claimed to give a proof of the frequency interpretation[15-18], but as we show in section 2.4 the proof is exact only if we take an *actually infinite* number of elements in the ensemble. If the number of elements is finite, as is always the case in all physical experiments, there are still doubts concerning the validity of the statement as a proposition that can be deduced from the other axioms of quantum theory (see the criticisms in[18]).

In the measurement of a property A (represented by the operator $\mathbf{A}$) of the quantum system in a pure state described by $|\Psi\rangle$ the eigenvalues a_n of $\mathbf{A}$ appear, in general, with different frequencies. The mean value of the observable A is calculated by,

$$\overline{A} \equiv \langle A \rangle = \langle \Psi | \mathbf{A} | \Psi \rangle. \tag{2.21}$$

If the conditions of the experiment do not fix the quantum state as a pure state, i.e., if we have different state vectors with different probabilities, then we have a mixture of states which, as already said, is described by a density matrix ρ. In this case the mean value of the observable A is given by,

$$\overline{A} = Tr(\rho \mathbf{A}). \tag{2.22}$$

2.2 THE WHOLE AND ITS PARTS

If we have a system composed of two parts, U and V, then the Hilbert space of the composed system is supposed to be given by the tensor product $\mathcal{H}^U \otimes \mathcal{H}^V$ of the Hilbert spaces $\mathcal{H}^U$ and $\mathcal{H}^V$ of the individual systems. Then, if U is in the state $|u\rangle$ and and V is in the state $|v\rangle$ the whole system can be described by the state vector[5]

$$|\Phi_{UV}\rangle = |u\rangle \otimes |v\rangle. \tag{2.23}$$

More generally, if U can be in principle in any of the states $\{|u_i\rangle\}, i = 1, 2...$ which are an orthonormal basis of $\mathcal{H}^U$, and V can be in principle in any one of the states $\{|v_j\rangle\}, j = 1, 2...$ which are an orthonormal basis of $\mathcal{H}^V$, than the whole system can be found in the state

$$|\Psi_{UV}\rangle = \sum_{i,j} c_{ij}|u_i\rangle \otimes |v_j\rangle. \quad (2.24)$$

The following question cannot be avoided: If the whole is described by $|\Psi_{UV}\rangle$, what is the state of its part U (or V)?

If $\mathbf{A} = \mathbf{1}_U \otimes \mathbf{A}_V$ is the representative of some observable A_V of the subsystem V, then the mean value of $\mathbf{A}$ in the state $|\Psi_{UV}\rangle$ is

$$\overline{A} = \langle\Psi_{UV}|\mathbf{A}|\Psi_{UV}\rangle = \sum_{ij,rs} c_{ij}^* c_{rs} \langle u_i|\mathbf{1}_U|u_r\rangle\langle v_j|\mathbf{A}_V|v_s\rangle =$$

$$\sum_{ij,rs} c_{ij}^* c_{rs} \langle u_i|u_r\rangle\langle v_j|\mathbf{A}_V|v_s\rangle = \sum_{ijs} c_{ij}^\dagger c_{is} \langle v_j|\mathbf{A}_V|v_s\rangle. \quad (2.25)$$

We can also calculate $\overline{A}$ in terms of the density matrix $\mathbf{1}_U \otimes \rho_V$ of the subsystem V, where ρ_V is defined as follows. If

$$\rho = \sum_{ijrs} |u_r\rangle \otimes |v_s\rangle \rho_{rs,ij} \langle v_j| \otimes \langle u_i| \; ; \rho_{rs,ij} = c_{ij}^* c_{rs} \quad (2.26)$$

is the density matrix of the whole, then

$$\rho_V = Tr^{(U)} \rho = \sum_t \langle u_t|\rho|u_t\rangle \quad (2.27)$$

i.e., we average over the system U. Then, calculating $\overline{A} = Tr(\rho_V \mathbf{A})$ we get like in Eq. (2.25)

$$\overline{A} = Tr(\rho_V \mathbf{A}) = \sum_{ijs} c_{ij}^\dagger c_{is} \langle v_j|\mathbf{A}_V|v_s\rangle. \quad (2.28)$$

But then, it follows that $Tr\rho_V^2 \leq 1$. The equality holds only in the special case when the state vector of the system is the product of the state vectors of the parts. Since eq.(2.28) holds we are tempted to ask : can we interpret the density matrix ρ_V of the subsystem V in terms of a mixture of states? We show now that such an interpretation is impossible!

Indeed, if the interpretation in terms of a mixture holds then there must exist some density matrix $\rho_{\alpha,m}$ such that for the observables A and B of the systems U and V, represented by the operators

$$\mathbf{A} = \mathbf{A}_U \otimes \mathbf{1}_V = |u_\beta\rangle\langle u_\beta| \otimes \mathbf{1}_V; \; \mathbf{B} = \mathbf{1}_U \otimes \mathbf{B}_V = \mathbf{1}_U \otimes |v_n\rangle\langle v_n| \quad (2.29)$$

we have respectively the probabilities $\delta_{\alpha\beta}$ and δ_{mn}. Obviously,

$$\rho_{\alpha,m} = |u_\alpha\rangle \otimes |v_\beta\rangle\langle v_\beta| \otimes \langle u_\alpha| \quad (2.30)$$

But then the whole system, composed of subsystems U and V is described by the density matrix

$$\rho' = \sum_{\alpha,m} p_{\alpha m} \rho_{\alpha,m} \qquad (2.31)$$

where the $p_{\alpha m}$ are the corresponding weights. But ρ' is, of course, the description of a mixture and not of the pure state $|\Psi_{UV}\rangle$ to which the density matrix ρ given by Eq. (2.26) corresponds.

The situation can be explained more easily as follows. If the subsystem U is with some probability in the state $|u_1\rangle$, with another probability in the state $|u_2\rangle$, etc..., then the whole must be with some probability in some state, but with another probability in another state, etc..., and this means that it cannot be in one and the same pure state. This shows that the only possible interpretation that we can give to the each eigenvalue of the density matrix ρ_V is not as *probability for the subsystem to be in some pure state* but as *probability of a transition to a pure state* if we take a measurement in the subsystem.

So, there is a clash between the whole and its parts in quantum physics. If we prepare the whole in some pure state and then measure observables for the parts, the whole will change its state and shall be described either by a new pure state or a mixture of states depending on if the observer registers or not the result of his (or her) measurements for the parts. Sometimes this is called *complementarity of the whole and its parts*. The whole has some features of *wholeness* which cannot be obtained from its parts!

Measurements play a crucial role in understanding *leaps* of the state vectors, when dealing with properties of the whole and its parts. If the whole were measured as existing in some definite state it is in general impossible to say that the parts are also in some definite states. But, if the parts are measured then they move into definite states.

2.3 CLASSICAL LIMIT AND DECOHERENCE

It is natural to think that we must obtain classical physics with its laws as some approximation in quantum physics. Of course, such a view is based on the belief that there is only one ultimate reality which is the quantum world and that classical physics is a consequence of quantum theory.

However, differently from going from relativistic physics to the non-relativistic domain by taking (in an appropriate way) the limit $v/c \to 0$, where c is the velocity of light in vacuum, the situation concerning how to go from quantum to classical theory is not so clear. In the extreme, there is a view of a *two-worlds picture*[6]: there is a world of *quantum phenomena* and a world of *classical phenomena* and neither of them is preferable or more ultimate than the other.

Nevertheless, let us discuss very briefly some *ways* of going from quantum to classical physics. Most of these *ways* concern the Schrödinger equation and evolution due to it and try to show how classical deterministic laws of motion follow from the quantum law of evolution.

This was the textbook approach until a few years ago when changes due to *measurement* (and the consequent wave function collapse) was not discussed very much.

However, in literature we can find many important discussions on the problem of the *appearance of quantum leaps* in classical physics (Schrödinger's cat paradox) and on the possible *ways* to solve the problem, discussed in the subsubsections below.

2.3.1 EHRENFEST THEOREM

From Schrödinger equation for the wave function (or Heisenberg evolution for observables) we can obtain Newton's second law of motion:

$$\frac{d}{dt}\vec{r} = \frac{1}{m}\vec{p}, \quad \frac{d}{dt}\vec{p} = \vec{F}, \quad (2.32)$$

where,

$$\vec{r} = \int d^3x \Psi^*(\vec{x}) \mathbf{x} \Psi(\vec{x}),$$

$$\vec{p} = \frac{\hbar}{i} \int d^3x \Psi^*(\vec{x}) \nabla \Psi(\vec{x}),$$

$$\vec{F} = -\int d^3x \Psi^*(\vec{x}) [\nabla V(\vec{x})] \Psi(\vec{x}). \quad (2.33)$$

In Eq. (2.33) $V(\vec{x})$ is the potential function. From the above formulas we may have the impression that classical physics does indeed follow from quantum theory in a way similar to that in which thermodynamics follows from statistical physics, i.e, by making averages on ensembles. However, even at this level we can see some problems. For example, Newton's law can be obtained only for some very special potentials and wave functions. It is trivial to obtain it for $\nabla V(\vec{x}) = const.$, but in the general case we usually think that the wave function is so localized that it is impossible to consider $\nabla V(\vec{x}) = const.$, over distances where $\Psi(\vec{x})$ is changing.

Ehrenfest's theorem can explain some situations like the behavior of electrons in cathode ray tubes, while in other situations there are problems, because the force $\vec{F}$ in Eq. (2.33) is not identical to the Newtonian force $-\nabla V(\vec{x})$. Even for the movement of the Moon it is not possible to understand what the meaning of a *localized* wave function is. Is it possible for the Moon's wave function to become delocalized? The de Broglie wave length for the motion of the Moon's center of mass (if we take its Compton wavelength) is much smaller than its gravitational radius. So, if we could have such a localized wave function we could see Hawking's explosion of the macro object as a *mini-black hole* (see [19]).

Ehrenfest's theorem shows that there are two important features necessary in order to obtain classical physics: the first is by utilizing averages and the second is by having a special dynamic law of evolution.

2.3.2 CLASSICAL DYNAMICAL VARIABLES AND COLLECTIVE OBSERVABLES

Von Neumann suggested the idea that in classical physics we measure some special, always *commuting macroobservables,* such that the eigenvalues for these macroobservables are expectation values for the noncommuting observables. For the macroobservables we can then construct the *phase space* where classical physics is valid. Another name for the macroobservables may be *collective observables* because they correspond in general to measuring some properties of an ensemble of microparticles. However, it is important to keep in mind that there are situations where the collective observables are not described by classical physics, as, is the case, e.g., of the macroscopic quantum mechanics of superfluids, superconductivity, etc...

According to Omnès[20], given a macroscopic classical object, it can be represented by a family of states corresponding to a subspace $\mathcal{H}_s$ in the Hilbert space $\mathcal{H}$ of the system of particles from which it is made. We define the *collective position observables* as a set of commuting observables acting on $\mathcal{H}_s$. They are almost equal to the variables used in the formulation of classical mechanics. For a pendulum, e.g., they specify its position, but we can include if necessary, other collective variables like elastic deformations, the wire, etc. In principle it is even possible to consider atom positions as collective variables, by using, e.g., the Born-Oppenheimer approximation. We can then complete the collective position observables by other observables in order to get a maximal set of commuting collective observables. To apply the quantum theory to a macroscopic classical object, it is fundamental to interpret it as a *formal physical system* being made up of two interacting physical systems, called the *collective system* and the *environment*. The states of a collective system with n collective degrees of freedom are associated with wave functions $\langle \vec{q} | \psi \rangle = \psi(\vec{q})$, $|\psi\rangle \in \mathcal{H}_s$, where $\vec{q} = (q^1, q^2, ..., q^n)$ are the coordinates of the configuration space $\mathcal{M}_q$ of the collective system. These wave functions are square integrable with a scalar product relative to a weight function $\mu(\vec{q})$,

$$\langle \psi_1 | \psi_2 \rangle = \int_{\mathcal{M}_q} d\vec{q}\, \mu(\vec{q}) \psi_1^*(\vec{q}) \psi_2(\vec{q}) \tag{2.34}$$

A collective Hamiltonian $\mathbf{H}_c$ is associated to the collective system and the weight function $\mu(\vec{q})$ is chosen in such a way that the scalar product is invariant under time evolution given by the Schrödinger equation with the collective Hamiltonian.

Microscopic coordinates are associated with the environment, which is further divided in *internal environment* and *external environment*. The internal environment consists of the matter from which the classical object is made. The external environment is the medium where the classical system is.

The total Hamiltonian of the collective system plus environment is made up of three contributions,

$$\mathbf{H} = \mathbf{H}_c + \mathbf{H}_e + \mathbf{H}_{int} \tag{2.35}$$

where $\mathbf{H}_c$ depends only on the collective variables; $\mathbf{H}_e$ does not depend on them, and represents internal energy. $\mathbf{H}_{int}$ is responsible for *dissipation*, since it couples the collective system and the environment, permitting the exchange of energy between both systems.

A *macro-observable* can be defined by a self-adjoint operator $\mathbf{M}$ acting on the Hilbert space $\mathcal{H}$ of the collective system (see Omnès[20]).[7] To relate $\mathbf{M}$ with a classical dynamical variable[8] $m(\vec{q}, \vec{p})$ we first recall that Wigner[21] in 1932 proposed a quantum analog to the Liouville function $f(\vec{q}, \vec{p})$ of statistical mechanics. For a quantum system described by the density matrix $\rho(\vec{q}, \vec{q}\,')$ the Wigner function is defined by

$$f_W(\vec{q}, \vec{p}) = \frac{1}{(\pi\hbar)^n} \int d\vec{r} \rho(\vec{q} - \vec{r}, \vec{q} + \vec{r}) e^{2i\vec{p}\cdot\vec{r}} \tag{2.36}$$

For a pure state $\psi(\vec{q})$, with Fourier transform $\psi(\vec{p})$, we have

$$\rho(\vec{q}, \vec{q}\,') = \psi(\vec{q})\psi^*(\vec{q}\,'), \quad \tilde{\rho}(\vec{p}, \vec{p}\,') = \psi(\vec{p})\psi^*(\vec{p}\,') \tag{2.37}$$

and we see that Eq. (2.36) gives the right marginal distributions

$$\int d\vec{p} f_W(\vec{q}, \vec{p}) = \rho(\vec{q}, \vec{q}), \quad \int d\vec{q} f_W(\vec{q}, \vec{p}) = \tilde{\rho}(\vec{p}, \vec{p}) \tag{2.38}$$

From Eq. (2.36) we immediately get for two different density matrices ρ_1 and ρ_2

$$\int\int d\vec{q} d\vec{p} f_W^1(\vec{q}, \vec{p}) f_W^2(\vec{q}, \vec{p}) = \frac{1}{(2\pi\hbar)^n} Tr(\rho_1 \rho_2)$$

$$\int\int d\vec{q} d\vec{p}\, [f_W(\vec{q}, \vec{p})]^2 \leq \frac{1}{(2\pi\hbar)^n} \tag{2.39}$$

with the equality in the second Eq. (2.39) holding only for pure states. Since for pure states $Tr(\rho_1 \rho_2) = |\langle \psi_1 | \psi_2 \rangle|^2$ it follows that the Wigner functions of orthogonal states satisfy

$$\int\int d\vec{q} d\vec{p} f_W^1(\vec{q}, \vec{p}) f_W^2(\vec{q}, \vec{p}) = 0 \tag{2.40}$$

Equation (2.40) shows that the f_W's must be negative sometimes and for this reason they cannot be interpreted as probability distributions, despite the analogy with the Liouville densities, which becomes obvious since for any two functions $g_1(\vec{q})$ and $g_2(\vec{p})$ using Eq. (2.37) we have

$$\langle g_1(\vec{q}) + g_2(\vec{p}) \rangle = \int\int d\vec{q} d\vec{p} f_W(\vec{q}, \vec{p})\, [g_1(\vec{q}) + g_2(\vec{p})] \tag{2.41}$$

as is the case in classical statistical mechanics. However, as pointed out by Moyal[23], no such formula holds for more general functions of $\vec{q}$ and $\vec{p}$ due to factor ordering ambiguities. Only for polynomial functions can a unique ordering be defined. It is also important to remark that $f_W(\vec{q}, \vec{p})$ does not tend to a definite limit when $\hbar \to 0$, but presents instead increasing oscillations.

Despite all these limitations the Wigner functions give a qualitative visualization of the dynamic behavior of a quantum system with n degrees of freedom. Due to recent advances in mathematics, the so called Weyl calculus, a subbranch of *microlocal analysis* or *pseudo-differential calculus*[24,25], there exists now a rigorous theory of how to associate to a given operator **M** (associate to a quantum system with n degrees of freedom) a real function $m(\vec{q}, \vec{p})$ in phase space called its *symbol*. Inspired by the theory of the Wigner function such a correspondence is given in the case where the configuration space is Euclidean, by[20]

$$m(\vec{r},\vec{p}) = \iint d\vec{r}\,'d\vec{r}\,''\langle\vec{r}|\mathbf{M}|\vec{r}\,''\rangle \delta\left[\vec{r} - \frac{(\vec{r}\,' + \vec{r}\,'')}{2}\right] \exp\left[\frac{i}{\hbar}\vec{p}\cdot(\vec{r}\,'' - \vec{r}\,')\right] \tag{2.42}$$

which can be called the *Wigner formula*.

The correspondence between m and $\mathbf{M}$ is one to one and satisfies the following conditions:

(i) It is linear.

(ii) $\mathbf{q}^i$ and $\mathbf{p}_i$, the position and the corresponding momentum observables are associated with the phase space coordinates (q^i, p_i).

(iii) Let $\mathbf{F}(\mathbf{q},\mathbf{p})$ be a Hermitian operator that is a homogeneous second order polynomial, then $\mathbf{U} = \exp[i\mathbf{F}(\mathbf{q},\mathbf{p})]$ is a unitary transformation and transforms an observable $\mathbf{M}$ into $\mathbf{M}' = \mathbf{U}^{-1}\mathbf{M}\mathbf{U}$. These transformations make the observables $(\mathbf{q}',\mathbf{p}')$ to be related with the observables $(\mathbf{q},\mathbf{p})$ by a linear canonical transformation, and are the simplest transformations permitting a correspondence between unitary and canonical transformations. Equation (2.42) implies that the symbol m' associated with $\mathbf{M}'$ can be obtained by using the same canonical transformation upon the variables $(\vec{q},\vec{p})$ in the function $m(\vec{q},\vec{p})$.

(iv) $m(\vec{q},\vec{p})$ is real if $\mathbf{M}$ is Hermitian.

In statistical classical mechanics the coordinates $(\vec{q},\vec{p})$ in phase space for a *non-chaotic* system with n degrees of freedom lie in some connected and simply connected domain $C(t)$ that changes with time. Omnès[20] studied the following important question: can we give a meaning in quantum theory to this classical property, i.e., can we associate a projector in the Hilbert space of a collective wave function with a classical property? Omnès showed that we can associate *quasi-projectors,* which are defined as self-adjoint operators having only discrete eigenvalues in the interval $[0, 1]$. A quasi-projector $\mathbf{F}$ of *rank* N and order η is defined by

$$Tr\mathbf{F} = N \tag{2.43}$$
$$Tr(\mathbf{F} - \mathbf{F}^2) = N\,O(\eta) \tag{2.44}$$

where $\eta = \varepsilon\theta$ is some *effective classicity parameter* which is much smaller than unity. If we have a connected and simply connected cell in phase space, with L and P as characteristic scales for the coordinates and momenta, and if $V(C)$ is the volume of the cell and $\Sigma(S)$ is the area of its boundary $S = \partial C$, then

$$\varepsilon = \left(\frac{\hbar}{LP}\right)^{\frac{1}{2}} ; \theta = \frac{\Sigma(S)}{V(C)} \tag{2.45}$$

It is well known that we can associate only one quantum state with a rectangular box in phase space having volume $(\hbar^n)$. A projector is more or less well defined if there are many such boxes in C, and the *rank* of the projector is defined by the number of semiclassical states,

$$N = \frac{1}{(2\pi)^n} \int_C \hbar^{-n} d\vec{q}\,d\vec{p} \tag{2.46}$$

To relate **F** with the cell C we consider a wave function $\psi_1(\vec{q})$ with averages for position and momentum $(\vec{q}_0 + \Delta\vec{q}, \vec{p}_0 + \Delta\vec{p})$. We say that $\psi_1(\vec{q})$ is *well inside* C when the $2n$-dimensional box has its center at $(\vec{q}_0, \vec{p}_0)$ and half-sides $(\Delta\vec{q}, \Delta\vec{p})$ is well inside C. Analogously, let $\psi_2(\vec{q})$ be a state *well outside* C. The quasi-projector **F** is associated with the cell C if

$$\mathbf{F}\psi_1(\vec{q}) \simeq \psi_1(\vec{q}), \quad \mathbf{F}\psi_2(\vec{q}) \simeq 0 \qquad (2.47)$$

With this convention the case of quasi-classical states corresponds to eigenstates of **F**. There are two different ways to associate **F** with a given cell. The first uses coherent states[9] $|g_{\vec{q},\vec{p}}\rangle$ and we have

$$\mathbf{F} = \int_C d\vec{q}d\vec{p}\,|g_{\vec{q},\vec{p}}\rangle\langle g_{\vec{q},\vec{p}}| \qquad (2.48)$$

The second way associates **F** with its symbol $f(\vec{q},\vec{p})$ given by the Wigner formula (Eq. 2.41), which is essentially the characteristic function of the cell, i.e., it is 1 in C and 0 outside, and must be smoothed down in the proximity of the boundary of C upon a region of width ε, for otherwise the operator is not well-behaved.

In order for the cell C to have a classical evolution, two conditions are necessary: firstly, it is necessary that the classical property associated with the initial cell C_0 must be valid for the initial state (which is not the case for the macroscopic quantum object) and secondly it is necessary that the deformation of the cell C_0 to $C(t)$ not be very large. The second condition rules out *chaotic systems* when the parameter θ increases exponentially in time[20].

By introducing a non-dimensional metric in phase space $\mathcal{M}_{qp}$ by[10]:

$$g_e : \sec T_e \mathcal{M}_{qp} \times \sec T_e \mathcal{M}_{qp} \to \mathcal{R}, \; e \in \mathcal{M}_{qp}$$

$$g = \frac{1}{L^2}\sum_i dq^i \otimes dq^i + \frac{1}{P^2}\sum_i dp_i \otimes dp_i \qquad (2.49)$$

it becomes possible to compare quasi-projectors that are associated with clearly separated cells C and C' of similar sizes. If δ is the minimum distance between C and C' in the metric (2.49) then the trace norm of the products of the associated quasi-projectors **F** and **F**' satisfy

$$Tr|\mathbf{FF'}| \simeq Tr|\mathbf{F'F}| \simeq O(N\exp(-\frac{\delta}{\varepsilon})) \qquad (2.50)$$

as it must be the case if they are independent. Then, quasi-projectors associated with two well separated classical properties satisfy Eq. (2.50) with high precision. It follows that it is possible using the language of quantum theory, to express the fact that the corresponding classical properties are well separated.

2.3.3 THE WKB APPROXIMATION

A second *way* to obtain classical physics from quantum theory for a given quantum system is given by situations where the *WKB* approximation, which is obtained from the Schrödinger equation by taking the limit $\hbar \to 0$, is a good one. As an example, consider some special *quasi-classical* wave function for a system constituted of a particle in an external field U

$$\langle \vec{x}|\Psi\rangle = \Psi(t,\vec{x}) = a(t,\vec{x})\exp[\frac{i}{\hbar}S(t,\vec{x})] \tag{2.51}$$

where a is a slowly varying real function of the coordinates. Then, from the Schrödinger equation, and neglecting the term $\frac{\hbar^2}{2ma}\nabla^2 a$, we get

$$\frac{\partial S}{\partial t} + \frac{1}{2m}(\nabla S)^2 + U = 0 \tag{2.52}$$

$$\frac{\partial a}{\partial t} + \frac{a}{2m}\nabla^2 S + \frac{1}{m}\nabla S \cdot \nabla a = 0 \tag{2.53}$$

Equation (2.52) is the classical *Hamilton-Jacobi equation* and Eq. (2.53), after multiplication for $2a$ is recognized as the *continuity equation*,

$$\frac{\partial a^2}{\partial t} + \nabla \cdot (a^2 \frac{\nabla S}{m}) = 0 \tag{2.54}$$

In these formulas $a^2 = \left|\left|\Psi\right\rangle\right|^2$ is the probability density and $\nabla S/m = \vec{p}/m = \vec{v}$ is the classical velocity of the particle.

For a Bose gas Eq. (2.54) is the real equation of continuity for the *condensate* of particles in the same state $|\Psi\rangle$ and a^2 is then the real density of the medium-condensate. There is a connection between what we discussed previously concerning Ehrenfest theorem and macro-observables, and the special form of the wave function discussed here. It can be shown that quasi-classical wave functions are eigenfunctions of commuting macro-observables[20].

Widely used as quasi-classical wave functions are also the *coherent states* already mentioned. In any case we see that in the classical limit somehow only special commuting operators survive and also only special wave functions from the whole Hilbert space can be used. We can also remember Bohr's idea that for large quantum numbers we should go from the quantum to the classical description, because the action becomes larger and larger than the Planck constant. This idea has now been tested in experiments with highly excited states of atoms, for which even the "classical orbits" for the electrons become visible[27]. However, the electron's motion is not totally classical, as is the case of the motion of planets around the Sun, and indeed what is observed is a kind of *intermediate* regime, mixing quantum and classical features. This leads us to the next *way* and the *crucial* problem: how noncommuting observables disappear in the classical limit.

2.3.4 DECOHERENCE AND SCHRÖDINGER'S CAT PARADOX

We discuss now the so-called *Schrödinger's cat paradox*, as an example of the problem of disappearance of interference effects for macrobodies (*decoherence*) and the impossibility of measuring some observables. This leads to a *breakdown of the superposition principle* in quantum theory and the impossibility of the realization of some wave functions for macrobodies.

Attempts to solve this problem, which can be put as: *Are trees, cats, trucks, etc..., described by de Broglie waves?* can be classified in two groups.

(i) Decoherence due to the environment. This idea is due to Feynman, Vernon[28] and Zurek[29]. Let us take as example a pendulum, which is described as a collective system in interaction with the environment. According to these authors, decoherence is a dynamic effect of disappearance of interference effects due to the loss of local phase correlations for wave functions of the environment. For the pendulum, in particular, we have an *internal* environment, which is the *formal physical system* made up of the particles of the pendulum after the separation of the center of mass, and an *external* environment, composed of, e.g., the air molecules and photons around the pendulum. Let us suppose that the internal environment is initially in thermal equilibrium at zero temperature, such that the environment is in the ground state.

Following Omnès[20], let the pendulum be released from a position x_1, or x_2 with zero velocity. The initial state of the pendulum plus environment corresponding to each one of the definite possibilities above can then be represented by

$$|\Psi_i\rangle = |\psi_i\rangle \otimes |o\rangle \in \mathcal{H}_c \otimes \mathcal{H}_e, \ i = 1, 2 \qquad (2.55)$$

where $|\psi_i\rangle$ is the state of the collective system such that $\psi_i(x) = \langle x|\psi_i\rangle$ has respectively the average values x_i ($i = 1, 2$) and zero for the position and momentum. $|o\rangle$ is the the ground state of the environment. Now, consider the following superposition for the total system

$$|\Psi(t)\rangle = a(t)|\Psi_1\rangle + b(t)|\Psi_2\rangle \qquad (2.56)$$

The aim is to obtain from this superposition the *mixture* of states for which interference effects are not observable. As we know, the probabilities of particular results for a given observable associated with the collective system at some instant t are calculated using the reduced or collective density operator $\rho_r(t)$, which is obtained from $\rho(t) = |\Psi(t)\rangle\langle\Psi(t)|$ by taking the partial trace over the environment variables. The crucial point in the whole idea is to consider an interaction Hamiltonian $\mathbf{H}_{int}$, coupling the collective system to the environment such that it leads to a dissipation.

Caldeira and Leggett[30] obtained such a dissipation effect by taking $\mathbf{H}_{int}$ in a special form, linear in the positions of the environment oscillators and not involving their momenta. Then, since the wave functions of the environment are sensitive to the collective motions, they become orthogonal very rapidly and as a consequence the non-diagonal terms of the reduced density matrix also vanish rapidly. Then, the authors of this *way* make a *strong claim*, treating the coefficients $|a(t)|^2$ and $|b(t)|^2$ of the diagonal reduced density matrix as *being* the probabilities of the system to be in the states $|\Psi_1\rangle$ or $|\Psi_2\rangle$. Since decoherence has the same physical origin as dissipation, it depends on the dissipation coefficient λ that appears in the average collective motion of the pendulum, which, as well known is described by the following classical equation,

$$\frac{d^2x}{dt^2} + \lambda\frac{dx}{dt} + \omega^2 x = 0 \qquad (2.57)$$

where $x(t)$ is the collective variable giving the position of the pendulum, and ω is the frequency.

The factor $I(t)$ in the off diagonal elements of the reduced density matrix calculated with the Caldeira and Leggett interaction Hamiltonian is

$$I(t) = \exp\{-\frac{1}{4\hbar}m\omega(x_1 - x_2)^2[1 - \exp(-\lambda t)]\} \qquad (2.58)$$

Taking the mass of the pendulum as $m = 1(\text{g})$, the period $2\pi/\omega = 1(\text{sec})$ and the damping time $1/\lambda = 60(\text{sec})$, we find for an initial distance $(x_1 - x_2) \approx 1$ micron that after $t = 1$ (nanosecond), $I(t) = \exp(-10^5)$!

Another example is given by the motion of the Moon. Its center of mass is decoupled from its *internal* environment but there is interaction with (among other things) the light coming from the sun, which constitutes the *external* environment here. The exponential decrease $\exp(-t/T)$ in this case can be calculated with the Caldeira and Leggett theory, with $T \approx GNc$ (where G is the Moon cross section for the scattering of photons coming from the sun, N is the photon density and c is the velocity of light), and leads to the decoherence effect for $t \approx 10^{-35}$ sec.

In spite of these beautiful examples, *surely* we do not have here a clear answer to our question, namely, how the noncommuting observables disappear in the classical limit.This is indeed the case because, as we have already explained, it is *impossible* to interpret the diagonal terms of the reduced matrix in terms of a *mixture of states*, for we then arrive at a contradiction with the assumption of a pure state of the whole system.

What these examples of *decoherence* really show is that for macrobodies there is some set of *preferable observables*, such that if they (the collective variables) are measured for a sub-system (here, the collective system) then the evolution in time of the whole system leads to decoherence. Now, if we measure some other observables, could we see interference effects? The usual *answer* is yes, but with the addendum that it is difficult, in practice, to measure these other observables, because in order to do so it is necessary to take into account correlations for a great number (Avogadro's number) of microparticles. This *practical answer* is not an answer about Nature, but only about our human possibilities.

The *big question about Nature* is then: are all non-commuting observables for macrobodies (as systems of a great number of particles) *observables*?

If the answer is yes (and quantum theory is indeed a valid theory in all possible circumstances), this means that we, human beings, due to our special biological situation, measure and see only part of the real quantum universe described by quantum theory. This is a kind of *anthropic* reasoning about the Universe. Our bodies are such that we can measure only special observables consistent with our biological survival. It may be that measurement of non-commuting observables for the macroworld is inconsistent with life and is only for *very distant* objects, like those in the microworld or in the cosmos, that noncommuting observables can be measured without danger to life. This doesn't mean that in the future, by using special techniques, that permit making such measurements harmlessly, human beings will not measure non-commuting macro-observables.

Now, we discuss briefly the other *way* to get commuting observables.

(ii) Classical limit as going to a system with infinite number of degrees of freedom. In the algebraic axiomatic approach to quantum theory[31] it is shown that for a system with an infinite number of degrees of freedom, differently from the case of finite systems, new observables arise, and commute with all elements of the noncommutative algebra of local observables. For irreducible representations of the algebra of observables, these operators, by Schur's lemma are multiples of the identity operator, with values depending on the representation. Classical observables, in this view are similar to intensive observables of thermodynamics, like, e.g., particle density, energy density, etc. Hepp[32] showed that for a system with infinite degrees of freedom it is possible to have an evolution in time where the pure state (which is a superposition of states) becomes a state which has the same properties as a mixture. But Hepp's analysis showed also that the time for this evolution is infinite! So here, too, there is the same problem as in the *decoherence effect*: the pure state is one state, mixture means many states, but "$1 \neq 2$", so logically there is still a problem.

Another problem of this *way* is actual infinity. For any finite system we do not get the above results and the fact is that we do not see infinite systems; it is only a mathematical trick.

All this shows that the great problem of *Schrödinger's cat,* or how non-commuting observables disappear in the classical limit, is still unsolved.

We have today a *two worlds picture.* All observables are observables for the microworld, but only some preferable observables are observables for macrobodies!

Do human observers or Nature have this preference?

2.4   PROOFS OF THE FREQUENCY INTERPRETATION

Everett[15], Finkelstein[16] and Hartle[17] discussed in their papers proofs of the *frequency interpretation* of the wave function, claiming that this interpretation is not some special postulate, but is a consequence of the mathematical formalism of quantum mechanics.

Let us analyze their proof following[17]. We consider a quantum system in the state $|S\rangle$ such that $\langle S|S\rangle = 1$. An observable A is measured. To it corresponds a self-conjugate operator $\mathbf{A}$ with eigenvectors $|i\rangle$ such that,

$$\mathbf{A}|i\rangle = a_i |i\rangle \tag{2.59}$$

Let us take an *ensemble* of N identical systems in the state $|S\rangle$. Suppose that it is possible to discriminate every system in the ensemble by some index α, $\{...|S,\alpha\rangle...\}$, $\alpha = 1, 2, ...N$. Supposing that the N different copies of the system do not interact, the state vector of the ensemble of the N equally prepared systems is

$$|S^N\rangle = |S,1\rangle \otimes |S,2\rangle \otimes ... \otimes |S,N\rangle \tag{2.60}$$

Now, let $N \to \infty$ and introduce the state vector of the ensemble with an infinite number of systems:

$$|S^\infty\rangle = |S,1\rangle \otimes |S,2\rangle \otimes ... \tag{2.61}$$

24 NONLOCALITY IN QUANTUM PHYSICS

The main conclusion of [16] is that the vector $|S^\infty\rangle$ is an eigenvector of some *special operator* whose eingenvalue has the meaning of the frequency of getting the result a_i if A is measured in the state $|S\rangle$.

This frequency is equal to $|\langle k|S\rangle|^2$ for $i = k$. The special operator is $\mathbf{f}^{(k)}$, the frequency with which the number a_k can be found in the infinite ensemble:

$$\mathbf{f}^{(k)}|S^\infty\rangle = |\langle k|S\rangle|^2 |S^\infty\rangle \tag{2.62}$$

where

$$\mathbf{f}^{(k)} = \lim_{N\to\infty} \mathbf{f}_N^{(k)} \tag{2.63}$$

$$= \lim_{N\to\infty} \sum_{i_1,i_2,\ldots i_N} |i_1,1\rangle \otimes \ldots |i_N,N\rangle \left(\frac{1}{N}\sum_{\alpha=1}^N \delta_{ki_\alpha}\right)\langle i_N,N|\otimes\ldots\langle i_1,1|$$

The terms in parentheses in Eq. (2.63) is the weight for the states $|i_1\rangle,\ldots|i_N\rangle$ to be in the state $|k\rangle$. The operator $\mathbf{f}_N^{(k)}$ acts on $|S^\infty\rangle$ as

$$\mathbf{f}^{(k)}|S^\infty\rangle = \mathbf{f}^{(k)}|S^N\rangle \otimes |S^{N+1}\rangle \otimes |S^{N+2}\rangle \otimes \ldots \tag{2.64}$$

Now, let us show (supposing the existence of the limit given by Eq. (2.64))[11] that

$$\|\mathbf{f}^{(k)}|S^\infty\rangle - |\langle k|S\rangle|^2 |S^\infty\rangle\| = \lim_{N\to\infty} \|\mathbf{f}_N^{(k)}|S^\infty\rangle - |\langle k|S\rangle|^2 |S^\infty\rangle\|. \tag{2.65}$$

Indeed, from Eq. (2.62) we can write Eq. (2.65) as

$$\|\mathbf{f}^{(k)}|S^\infty\rangle - |\langle k|S\rangle|^2 |S^\infty\rangle\| = \lim_{N\to\infty} \|\mathbf{f}_N^{(k)}|S^N\rangle - |\langle k|S\rangle|^2 |S^N\rangle\|_N \tag{2.66}$$

where $\|\ \|_N$ is the norm in the Hilbert space of an ensemble of N systems. From the definition of $\mathbf{f}_N^{(k)}$ and orthonormality of the states it follows that the square of the norm in Eq. (2.66) is,

$$\|\mathbf{f}^{(k)}|S^\infty\rangle - |\langle k|S\rangle|^2 |S^\infty\rangle\|^2 = \sum_{i_1,\ldots i_N}\left[\sum_{\alpha,\beta=1}^N \frac{1}{N^2}\delta_{ki_\alpha}\delta_{ki_\beta}\right.$$

$$\left. - 2|\langle k|S\rangle|^2 \sum_{\alpha=1}^N \frac{1}{N}\delta_{ki_\alpha} + |\langle k|S\rangle|^4\right]|\langle i_1|S\rangle|^4\ldots|\langle i_N|S^\infty\rangle|^4 \tag{2.67}$$

Now, let us do the summation over $i_1,\ldots,i_N$ using the completeness of the set $\{|i\rangle\}$ and the normalization of $|S\rangle$. Considering separately the cases $\alpha = \beta$ and $\alpha \neq \beta$ and writing the corresponding terms we get,

$$\|\mathbf{f}^{(k)}|S^\infty\rangle - |\langle k|S\rangle|^2 |S^\infty\rangle\|^2 = \lim_{N\to\infty}[N^{-1}|\langle k|S\rangle|^2 +$$
$$N^{-1}(N-1)|\langle k|S\rangle|^4 - 2|\langle k|S\rangle|^2 + |\langle k|S\rangle|^2] = 0. \tag{2.68}$$

It is then proved that if we take the infinite ensemble as one *individual system* with an infinite number of degrees of freedom, then the states $|S^\infty\rangle$ (for each $|S\rangle$) are eigenvectors of the operator of the frequency of observation of A with eigenvalue a_i.

Experimentally, the observed frequency of the eigenvalue a_i is the result of measurements made in a large number of systems all prepared in the same state. The index α that discriminates the different elements of the ensemble is in general the index of the instant of time at which the individual system is prepared. For example, if an electron is prepared in an electronic pump, then α means that at t_1 the electron is prepared in the state $|S\rangle$ and then A is measured. The same procedure is repeated at t_2, at t_3, etc...

The *criticism* of d'Espagnat[11] and others to the above proof is based on the observation that the proof is correct only for an actually infinite sequence of states.

Now, what do we have to say about a real sequence that can be only very large, but finite? Here, we are in the same position as with the von Mises statistical definition of probability. Why for finite sequences the frequencies of definite results are close to those given by quantum theory? Is each particular frequency *oscillating* close to the value given by the quantum rule?

The experimental support for quantum theory shows that it really does! But here there is no Bernoulli theorem showing that the sequences of small measures occur only very rarely. Quantum theory is not a standard probability theory; it uses probability amplitudes and there is no classical probability in it.

In standard probability theory when we claim, e.g., that the probability of a given coin that has been launched to be *up* is the same as to be *down* and that the frequency for each possibility oscillates around the value 1/2, an important role is played by the symmetry of the *object*, here the coin. In quantum theory, *symmetry* plays an even more important role (see, e.g.,Wigner [33]). But now, what is important is not the symmetry of the *object* but *symmetry itself as an object* (see Bohr and Ulfbeck[34]). This is the reason why the theory of the irreducible representations of the symmetry groups of quantum systems has a special status in the theory.

In Everett, DeWitt and Wheeler's [35] *many worlds interpretation* (MWI), it is possible to say that there really exist an infinite number of copies of our universe, so that there is no problem for finite sequences. Nevertheless, the *MWI*, which takes actual infinity as reality, has its own problems, which we shall discuss later. Ballentine[36], e.g., gave arguments showing that the *MWI* implies in equal probabilities 1/2 instead of $|\alpha|^2$ and $|\beta|^2$ for a spin 1/2 system in an arbitrary state $|\Psi\rangle = \alpha|\Psi_+\rangle + \beta|\Psi_-\rangle$.

In summary, we want to say that there are still many unsolved problems concerning the comparison of the probability concept in quantum theory and this concept in standard probability theory.

Notes

1. The postulate of wave function collapse (see **5** below) implies that two vectors $|\Psi\rangle$ and $e^{i\alpha}|\Psi\rangle$, $\alpha \in \mathcal{R}$ (the real field) which differ only by a phase factor, lead to the same physical predictions. Therefore an overall phase has no observable consequence for a system in a given state. Of course, for a state Ψ in a superposition as in Eq. (2.1) the relative phases are fundamental.

2. There is an exception in the application of the *PS* that has to do with so-called *super selection rules* for both the case of expansions like that in Eq. (2.1), which refers to observables with discrete spectra, as well as for the case of observables with continuous spectra. For example, a superposition of states corresponding to different values of the charge operator is certainly non observable and therefore is not a state of whatever physical system we can imagine.

3. See Appendix B for the definition of the conjugate dual space and of the linear operators acting on a given Hilbert space.

4. The spin space, as is well known is the carrier space of the $D^{\frac{1}{2}}$ representation of the group $SU(2)$ (the 2-dimensional special unitary group) which is the universal covering group of $SO(3)$ (the 3-dimensional special orthogonal group)[12].

5. Equation (2.23) is the correct description of the whole system only if U and V are distinguishable. If the parts are indistinguishable the correct wave function must take into account the statistics of the U and V systems.

6. Just as for the ancient Greeks the physical laws in the heavens were different from those on Earth.

7. Another possibility, which has been introduced by George, Prigogine and Rosenfeld (*GPR*)[19] in their study of large quantum systems is the *doubling* of the Hilbert space. If we have *microcoordinates* and *micromomenta* operators $\mathbf{r}_{mic}$ and $\mathbf{p}_{mic}$ that do not commute, we can obtain commuting *macrocoordinates* and *macromomenta* by taking the tensor product $\tilde{\mathcal{H}} = \mathcal{H} \otimes \mathcal{H}$, so that

$$\mathbf{r}_{mac} = \mathbf{r}_{mic} \otimes \mathbf{1}; \mathbf{p}_{mac} = \mathbf{1} \otimes \mathbf{p}_{mic}.$$

Using these ideas *GPR* obtained in many cases the characteristic features of macroscopic systems.

8. $(\vec{q}, \vec{p})$ represents the $2n$ local coordinates q^i and p_i of the phase space $\mathcal{M}_{qp}$ of a physical system with n degrees of freedom.

9. Coherent states are eigenfunctions of a combination of coordinate and momentum operators forming annihilation operators. See, e.g.,[26] for details.

10. In Eq. (2.49) $\sec T \mathcal{M}_{qp}$ means section of the the tangent bundle and $\mathcal{R}$ is the real field.

11. This result can indeed be proved. See[16].

References

[1] E. C. G. Stuckelberg, Quantum theory in the real Hilbert space, *Helv. Phys. Acta* **33**(8), 727–752 (1960).

[2] E. C. G. Stuckelberg and M. Guenin, Quantum theory in the Real Hilbert space. II, *Helv. Phys. Acta* **34**(9), 621–628 (1961).

[3] D. Finkelstein, J. M. Jauch, S. Schiminovich and D. Spliser, Quaternion representations of quantum groups, *J. Math. Phys.* **4**(1) 136–140 (1963).

[4] D. Finkelstein, J. M. Jauch, S. Schiminovich and D. Spliser, Foundations of quaternion quantum mechanics, *J. Math. Phys.* **3**(2) 207–220 (1962).

[5] D. Finkelstein, J. M. Jauch, S. Schiminovich and D. Spliser, Principle of general Q covariance, *J. Math. Phys.* **4**(6) 788–796 (1963).

[6] G. Emch, Mécanique quantique quaternionienne et relativité restreinte I, *Helv. Phys. Acta* **36**(6), 739–769 (1963).

[7] G. Emch, Mécanique quantique quaternionienne et relativité restreinte II, *Helv. Phys. Acta.* **36**(6), 770–788 (1963).

[8] S. Adler, *Quaternionic Quantum Mechanics and Quantum Fields* (Oxford Univ. Press, Oxford 1995).

[9] S. De Leo and W. A. Rodrigues, Jr., Quantum mechanics: from complex to complexified quaternions, *Int. J. Theor. Phys.* **36**(12), 2725–2757 (1997).

[10] P. Roman, *Some Modern Mathematics for Physicists and other Outsiders*, vol. 2 (Pergamon Press Inc., New York, 1975).

[11] B. d'Espagnat, *Conceptual Foundations of Quantum Mechanics* (W. A. Benjamin, Reading, MA, 1976).

[12] W. Miller, Jr., *Symmetry Groups and its Applications* (Academic Press, New York, 1972).

[13] J. von Neumann, *Mathematical Foundations of Quantum Mechanics* (Princeton University Press, Princeton, 1995).

[14] W. Feller, *An Introduction to Probability Theory and its Applications*, third edition (J. Wiley and Sons, Inc., 1968).

[15] H. Everett, III, 'Relative state' formulation of quantum mechanics, *Rev. Mod. Phys.* **29**,(3) 454–462 (1957).

[16] D. Finkelstein, The logic of quantum physics, *Trans. N. Y. Acad. Sci.* **25**, 621–637 (1962-63).

[17] J. B. Hartle, Quantum mechanics of individual systems, *Am. J. Phys.* **36**, 704–712 (1968).

[18] D. Deutsch, Quantum theory as a universal physical theory, *Int. J. Theor. Phys.* **24** (1), 1–41 (1985)

[19] A. A. Grib and V. Yu. Dorofeev, in: *Proceedings of the Fifth Seminar on Quantum Gravity*, edited by M. A. Markov, V. A. Berezin and V. P. Frolov (World Scientific Publishing Co., Singapore, 1990), p. 202–210.

[20] R. Omnès, Consistent interpretations of quantum mechanics, *Rev. Mod. Phys.* **64**(2), 339–382 (1994).

[21] E. P. Wigner, On the quantum correction for thermodynamical equilibrium, *Phys. Rev.* **40**, 749–759 (1932).

[22] C. George, I. Prigogine and L. Rosenfeld, The macroscopical level of quantum mechanics, *Kongoleske Danske Videnskabernes Selskab Matematisk-Fysike Meddelelser* **38**, 1–44 (1972).

[23] J. E. Moyal, Quantum mechanics as a statistical theory, *Proc. of the Cambridge Phil. Soc.* **45**, 99–124 (1949).

[24] M. Taylor, *Pseudo-Differential Operators* (Princeton University Press, Princeton, 1981).

[25] L. Hörmander, *The Analysis of Linear Partial Differential Operators* (Springer-Verlag, Berlin, 1985).

[26] R. J. Glauber, Coherent and incoherent states of the radiation field, *Phys. Rev.* **131**(6), 2766–2788 (1963).

[27] G. M. Nauenberg, C. Stroud and J. Yeazell, The classical limit of an atom, *Sci. Am.* **270**(6), 24–29 (1994).

[28] R. P. Feynman and F. L. Vernon, Jr., The theory of a general quantum system interacting with a linear dissipative system, *Ann. Phys. (N. Y.)* **24**, 118–173 (1963).

[29] W. H. Zurek, Pointer basis of quantum apparatus— into what mixture does the wave packet collapse, *Phys. Rev. D* **24**(6), 1516–1525 (1981).

[30] A. O. Caldeira and A. J. Leggett, Path integral approach to quantum Brownian motion, *Physica A* **121**(3), 587–616 (1983).

[31] P. Jordan, J. von Neumann and E. P. Wigner, On an algebraic generalization of the quantum mechanical formalism, *Ann. Math.* **35**, 29–64 (1934).

[32] K. Hepp, Quantum theory of measurement and macroscopic observables, *Helv. Phys. Acta* **45**(2), 237–248 (1972).

[33] E. P. Wigner, On the unitary representations of the Lorentz group, *Ann. Math.* **40**, 140–204 (1939).

[34] A. Bohr and O. Ulfbeck, Primary manifestation of symmetry-origin of quantal indeterminancy, *Rev. Mod. Phys.* **67**(1), 1–33 (1995).

[35] B. S. DeWitt and R. D. Graham (eds.), *The Many Worlds Interpretation of Quantum Mechanics* (Princeton University Press, Princeton, 1973).

[36] L. E. Ballentine, The statistical interpretation of quantum theory, *Rev. Mod. Phys.* **42**(4), 358–381 (1970).

3
HOW Ψ COLLAPSES

3.1 THE PROBLEM

The existence of two different ways of change of the state vector of a given quantum system seems strange to many physicists. Why must there be a special *cause* for change of the state vector, brought about by measurement or wave packet collapse? Isn't this *cause* just a normal physical interaction of the system with the apparatus, and if the apparatus consists of quantum particles why can't it be described by the Schrödinger equation?

For a system in a pure state the negative answer to this question has been given by von Neumann in his famous book[1]. Then, Wigner[2] showed that *measurements*, leaving the system plus apparatus in a state with a definite position of the *pointer*, cannot be described by the linear laws of quantum theory.

For measurements of the *second kind* (Landau measurements[3]), this result has been proved by d'Espagnat and here we follow his proof[4].

It is a very important proof, because sometimes people say: "The apparatus is a very complex system, consisting of a large number of quantum particles, so we are *ignorant* of the real state vector of the apparatus. Instead of the state vector we must then use some *mixture of states* described by some appropriated density matrix. Is it not the case that the whole density matrix for the measured system plus the the measuring apparatus evolves due to the Schrödinger equation, towards some new density matrix which can be interpreted as giving probabilities of definite results of the measurement?"

If this could be the case, then we could say that *probability* in quantum theory is as *ultimate* as in classical statistical mechanics and has *ignorance* as its explanation.

But the answer to this question is *negative*! Strange as it is, the *pointer* of the apparatus cannot have a definite position due to the interference terms if unitary evolution is valid.

3.2 COLLAPSE AND UNITARY EVOLUTION

Let be g the coordinate of some *pointer* of the apparatus ($\mathcal{A}$) (whose possible states belong to the Hilbert space $\mathcal{H}^\mathcal{A}$) and let **g** be the Hermitian operator corresponding to this observable, which we suppose here to have a discrete and non degenerate spectrum. Let g_n and $|F_n\rangle \in \mathcal{H}^\mathcal{A}$ be respectively the eigenvalues and eigenfunctions of **g**.

There is a one to one correspondence between the set $\{\,|F_i\rangle\,\}$ and the the *complete* set $\{|\psi_i\rangle\}$, where $|\psi_i\rangle \in \mathcal{H}^S$, ($\mathcal{H}^S \equiv$ Hilbert space of the system $\mathcal{S}$) are the eigenvectors of the operator **O** corresponding to the observable O that is to be determined by the apparatus $\mathcal{A}$. Let $|\Psi\rangle = \sum a_n |\psi_n\rangle$ be the initial state of the microsystem $\mathcal{S}$. The initial state of the whole system $\Sigma = \mathcal{S} + \mathcal{A}$ before measurement is then

$$|\Psi\rangle \otimes |F_0\rangle \in \mathcal{H}^S \otimes \mathcal{H}^\mathcal{A}. \tag{3.1}$$

Due to the interaction of $\mathcal{S}$ with $\mathcal{A}$ this wave function will evolve according to the Schrödinger equation. After the interaction has taken place $\mathcal{S}$ is not more in the state $|\Psi\rangle$ and we must write the new state function of Σ as

$$\sum_n |f_n\rangle \otimes |F_n\rangle \tag{3.2}$$

because $\{\,|F_i\rangle\,\}$ is a complete set.

Now, Landau and Lifshitz[3] claim that it is possible due to the "classical nature of the apparatus, to say that the system Σ is not described by the whole sum, but only by one term", namely

$$|f_n\rangle \otimes |F_n\rangle. \tag{3.3}$$

Writing

$$|f_n\rangle = a_n |\varphi_n\rangle \tag{3.4}$$

with $\langle \varphi_n|\varphi_n\rangle = 1$, from Eqs. (3.3) and (3.4) we can say that that φ_n is the normalized wave function of the system $\mathcal{S}$ *after* the measurement.

For non ideal measurements the $|\varphi_n\rangle$, however are not in general orthogonal. Even more, due to the perturbation of the system $\mathcal{S}$ by the apparatus they do not coincide with $|\psi_n\rangle$. However, the $|\psi_n\rangle$ do exist and form a complete orthogonal system. So, if $\mathcal{S}$ is in the eigenstate ψ_n of **O**, and if we measure g in $\mathcal{A}$, then g_n appears with probability equal to one, but the final state of the system Σ is

$$|\varphi_n\rangle \otimes |F_n\rangle. \tag{3.5}$$

If we do a measurement on the system $\mathcal{S}$ (which has interaction with the apparatus) and which is in the state

$$|\Psi\rangle = \sum_n a_n |\psi_n\rangle, \tag{3.6}$$

the probability of obtaining g_n for g is $|a_n|^2$ and according to Landau the state of Σ after the measurement is $|\varphi_n\rangle \otimes |F_n\rangle$.

If we consider an ensemble of identical initial states and in order for the statistical prediction of Eq. (3.6) to hold true concerning the g_n, it is necessary that the ensemble originally described by eq(3.1) be transformed in a mixture constituted by a proportion $|a_n|^2$ of composite systems in each state described by Eq. (3.5). But using the same arguments as the ones employed when we study *the whole and its parts* in section 2.2, it is easy to show that this is not possible, i.e., an ensemble cannot be represented at the same time by a pure state and a (proper) mixture, unless parts of the density matrix are not observable, which is not the case here.

Now, let us *generalize* the above situation by taking into account a *mixture of states*. Since we really do not know the initial state vector of the apparatus, the whole system Σ must be described by some density matrix ρ_Σ evolving in time according to the Schrödinger equation. Our question is: can the mixture ρ_Σ after unitary evolution be interpreted as a mixture of states with a definite position g_n for the pointer, and is this possible for any initial state $|\Psi\rangle$ of $\mathcal{S}$?

Let the initial state of the apparatus $\mathcal{A}$ be described by some density matrix $\rho_\mathcal{A}$. Diagonalizing this matrix, we obtain a description in terms of orthonormal state vectors $|F^{(\rho)}\rangle$ and p_ρ, the probabilities of the apparatus being in some $|F^{(\rho)}\rangle$. Then, if $\mathcal{S}$ is initially in some $|\psi_n\rangle$ (the same for an ensemble of apparatuses), then the whole system Σ is initially with probability p_ρ in the state

$$|\psi_n\rangle \otimes |F^{(\rho)}\rangle \tag{3.7}$$

After the interaction, this state definitively, according to Landau and Lifshitz, becomes

$$|\varphi_n\rangle \otimes |F_n^{(\rho)}\rangle \tag{3.8}$$

where $|F_n^{(\rho)}\rangle$ describes the state of the pointer and where $|\varphi_n\rangle \neq |\psi_n\rangle$. We can then say that after the interaction, the systems Σ in the ensemble are in the state given by Eq. (3.8) with probability p_ρ. Now, following Wigner, we see that $|F_n^{(\rho)}\rangle$ and $|F_{n'}^{(\rho')}\rangle$ are orthogonal for $n \neq n'$, because the positions of the pointers are different. But for $\rho \neq \rho'$, the vectors $|\varphi_n\rangle \otimes |F_n^{(\rho)}\rangle$ and $|\varphi_n\rangle \otimes |F_n^{(\rho')}\rangle$ are obtained by the same unitary transformation (time's evolution from $|\varphi_n\rangle \otimes |F^{(\rho)}\rangle$, $|\varphi_n\rangle \otimes |F^{(\rho')}\rangle$) which are orthogonal, i.e.,

$$\langle F_n^{(\rho)}|F_{n'}^{(\rho')}\rangle = \delta_{nn'}\delta_{\rho\rho'} \tag{3.9}$$

while $\{|F_n^{(\rho)}\rangle\}$ is not necessarily a complete set.

Now, take the state of $\mathcal{S}$ as the superposition given by Eq. (3.6). The ensemble of initial compound systems can be looked at as an ensemble of systems such that p_ρ-parts of it are in the state

$$|F_n^{(\rho)}\rangle \otimes \sum_n a_n|\psi_n\rangle = |\Upsilon_0^{(\rho)}\rangle. \tag{3.10}$$

From the linearity of the laws of quantum theory, it follows that the ensemble of final compound systems is constructed from systems in the states

$$\sum_n a_n |F_n^{(\rho)}\rangle \otimes |\varphi_n\rangle = |\Upsilon^{(\rho)}\rangle \qquad (3.11)$$

with probabilities p_ρ also. The question is then: can we identify this ensemble with an ensemble of states of the type

$$|Z^{(m,n)}\rangle = \sum_\rho X_\rho^{(m,n)} |F_n^{(\rho)}\rangle \otimes |\varphi_n\rangle \qquad (3.12)$$

i.e., with ensembles of states corresponding to a definite value g_n of the pointer and definite state vector of the quantum system?

We recall that the ensemble $\mathcal{E}$ described by the set of states $\{|v_\alpha\rangle\}$, with given probabilities, is identical to the ensemble $\mathcal{E}'$ described by some other set $\{|u_i\rangle\}$, if necessary and sufficient conditions are satisfied, such that the value of any observable of a given individual system averaged over both ensembles are equal. But then the u_i must be a linear combination of the $|v_\alpha\rangle$.

It follows that the answer to our question is positive if and only if the $|Z^{(m,k)}\rangle$ are linear combinations of the $|\Upsilon^{(\rho)}\rangle$, i.e., some coefficients $u_\rho^{(m,n)}$ exist such that,

$$|Z^{(m,k)}\rangle = \sum_\rho u_\rho^{(m,k)} |\Upsilon^{(\rho)}\rangle \qquad (3.13)$$

i.e.,

$$\sum_\rho X_\rho^{(m,k)} |F_m^{(\rho)}\rangle \otimes |\varphi_m\rangle = \sum_{n,\rho} u_\rho^{(m,k)} a_n |F_n^{(\rho)}\rangle \otimes |\varphi_n\rangle. \qquad (3.14)$$

Since by Eq. (3.9) the $|F_m^{(\rho)}\rangle$, $m = 1, 2...$ are linearly independent we get,

$$u_\rho^{(m,k)} a_n = \delta_{nk} X_\rho^{(m,k)} \qquad (3.15)$$

i.e.,

$$a_n = \delta_{nk} t_\rho^{(m,k)} \qquad (3.16)$$

with

$$t_\rho^{(m,k)} = \frac{X_\rho^{(m,k)}}{u_\rho^{(m,k)}} \qquad (3.17)$$

It is evident that Eq. (3.14) cannot be valid if more than one a_n is different from zero. It follows then that the ensemble— mixture of states given by Eq. (3.12) cannot be considered as the mixture of states given by Eq. (3.11). This means that a *definite position* for the pointer is not compatible with the laws of quantum theory.

Some physicists nevertheless have the opinion that the above difficulty is not that serious and might disappear if we lessen our desire for a definite position g_n of the pointer (with error Δg_n, if **g** has a continuous spectrum). These physicists prefer to say that *almost definitively* the pointer has a position $(g_n, g_n \pm \Delta g_n)$, with small probabilities for other positions.

However, even this *almost definite* position cannot be true. Indeed, to be true, it is necessary that the mixture of $|\Upsilon^{(\rho)}\rangle$ with probabilities p_ρ also be a mixture of states described by the state vectors,

$$|Z'^{(m,k)}\rangle = \sum_\rho X_\rho^{(m,k)} |F_m^{(\rho)}\rangle \otimes |\varphi_m\rangle + \sum_{\rho,\nu \neq m} Y_\rho^{(m,\nu,k)} |F_\nu^{(\rho)}\rangle \otimes |\varphi_\nu\rangle \quad (3.18)$$

with

$$Y_\rho^{(m,\nu,k)} \ll X_\rho^{(m,k)}. \quad (3.19)$$

The necessary condition for this to be true is $|Z'^{(m,k)}\rangle$ be a linear combination of the $|\Upsilon^{(\rho)}\rangle$, i.e.,

$$|Z'^{(m,k)}\rangle = \sum_\rho u_\rho^{(m,k)} |\Upsilon^{(\rho)}\rangle \quad (3.20)$$

But due to the linear independence of the $|F_m^{(\rho)}\rangle$ we have,

$$u_\rho^{(m,k)} a_m = X_\rho^{(m,k)} \quad (3.21)$$

$$u_\rho^{(m,k)} a_n = Y_\rho^{(m,\nu,k)} \quad (3.22)$$

for $n \neq m$.

From this it follows that

$$\frac{Y_\rho^{(m,\nu,k)}}{X_\rho^{(m,k)}} = \frac{a_n}{a_m} \quad (3.23)$$

and then Eq. (3.19) can be true if and only if one of the a_i is much smaller then the others, which is not the general case, e.g., for equal a_i, for any i it is not valid.

In conclusion, we proved that measurements cannot be described by the dynamic laws of quantum theory.

References

[1] J. von Neumann, *Mathematical Foundations of Quantum Mechanics* (Princeton University Press, Princeton, 1995).

[2] E. P. Wigner, The problem of measurement, *Am. J. Phys.* **31**(1), 6–15 (1963).

[3] L. D. Landau and E. M. Lifshitz, *Quantum Mechanics* (Pergamon Press, London, 1958).

[4] B. d'Espagnat, *Conceptual Foundations of Quantum Mechanics* (W. A. Benjamin, Reading, MA, 1976).

4
BELL'S INEQUALITIES

We now discuss Bell's famous inequalities and the consequences of their breakdown for quantum theory. Discovered by J. Bell in 1965[1], they led to the possibility of finding new inequalities and relations by Clauser, Horne, Shimony, Greenberger, Zeilinger and Hardy[2–5]. These results show the crucial difference between predictions of standard probability theory and the ones of quantum theory. Indeed, the breakdown of these inequalities can be verified experimentally and experiments (the most fascinating by Aspect et al[6,7]),[1] were realized to show the profound difference between quantum objects and anything which can be understood in classical terms.

In spite of the fact that many authors after Bell have found different forms of his inequalities, we shall call them *Bell's inequalities* for simplicity. As we are interested in conceptual problems, we shall begin with their simple forms discussed by d'Espagnat[8] and Stapp[9], and after which we discuss the original form found by Bell.

4.1 THE EINSTEIN, PODOLSKI AND ROSEN (EPR) ARGUMENT

EPR discussed in their paper[10] the idea that quantum objects exist as *elements of reality* in the same sense as classical objects. This means that they must have definite properties if there is no external disturbance due to a measuring apparatus. From this point of view, in quantum theory it is impossible to measure observables described by noncommuting operators not because these *properties* do not *exist* together at a fixed instant of time, but because complementary measuring apparatuses disturb the microsystem and prevent us from obtaining exact knowledge of it.

EPR discussed the possibility of checking their idea by an ingenuous way in which the *selected quantum system* is not disturbed. In order to achieve this goal we must have a two-particle system. The situation we are going to investigate is taken from Bohm[11],[2], where the two particles are spin 1/2 particles and are originally prepared in a singlet state. Suppose that we arrive at a situation where the particles' spacetime positions are such that their spacetime interval is space-like. Then, if in the laboratory frame (supposed inertial), we measure at a given time t the spin projection S_z for one of the particles and find it to be $+\hbar/2$ due to conservation of angular momentum, the spin projection S_z for the second particle must be $-\hbar/2$. In this experiment, the measuring apparatus does not produce any physical disturbance on the second particle if we believe the *special relativity dogma* that superluminal signals (influences) do not exist in nature.[3] But for the second particle it is possible to measure at the same time t the noncommuting spin projection S_x. Therefore, we can know through this experiment both S_x and S_z for both particles!

Now, the mathematical formalism of quantum theory says that noncommuting operators do not have common eigenvectors and so it is impossible to have a state where both properties *exist* together. *EPR* conclude then that quantum theory is not complete.

4.2 D'ESPAGNAT'S VERSION OF BELL'S INEQUALITIES

To check the *EPR* idea, now consider d'Espagnat's[8] version of Bell's inequalities. Suppose we have an object characterized by three properties A, B, C, to which correspond in quantum theory three noncommuting operators **A**, **B** and **C**.

Suppose, following *EPR*, that (contrary to quantum theory) the object has simultaneously definite values of A, B and C, which we suppose can take on the values ± 1.

Now, take an ensemble of identical objects and denote by A^+ the case where the value of A is $+1$ (same for B, C) and A^- the case where the value of A is -1 (same for B, C). Then, standard Boolean logic and set theory show immediately that

$$N(A^+B^-) = N(A^+B^-C^+) + N(A^+B^-C^-), \tag{4.1}$$

where $N(...)$ is the number of objects in the ensemble with the supposed properties. From the simple equalities

$$N(B^-C^+) = N(A^+B^-C^+) + N(A^-B^-C^+), \tag{4.2}$$
$$N(A^+C^-) = N(A^+B^+C^-) + N(A^+B^-C^-), \tag{4.3}$$

we immediately get the inequality

$$N(A^+B^-) \leq N(B^-C^+) + N(A^+C^-). \tag{4.4}$$

In order to check this experimentally, we must, following *EPR*, take *pairs of objects* in a fixed state, e.g., in a *singlet* state, where if one particle has B^-, the other has B^+, etc... Then, when we go from a number N of particles to $n(\ ,\ )$ pairs in which the first (second) particle possesses the property shown in the first (second) slot, we can easily find, for instance, that

$$n(A^+, B^+) \leq n(B^-, C^-) + n(A^+, C^+) \tag{4.5}$$

If the inequality (4.5) and the other similar ones are violated for quantum objects, we have the following possibilities.

(a) The properties of quantum objects do not *exist* as *elements of reality* in the *EPR* sense or are not *beables*, as Bell proposed to call the *EPR* claim[4].They *acquire* their definitive existence only if they are measured (Copenhagen interpretation) and then according to quantum theory they cannot *exist* simultaneously.

(b) There are superluminal influences which perturb particles separated by space-like intervals, implying in a breakdown of Einstein's special relativity. As already mentioned in footnote no. 3 above, eventually this possibility must be taken seriously in attention, in view of the fact that all relativistic wave equations have superluminal solutions and recent experimental results.

4.3 STAPP'S FORMULATION OF BELL'S INEQUALITIES

Let us suppose[9] that we have an object which has four properties A, B, A' and B', which are independent of each other and which take values ± 1. Our instruments measure these values. It is not difficult to verify that

$$AB + AB' + A'B - A'B' = \pm 2. \tag{4.6}$$

If we have N objects (indexed by r) and the values of their properties are randomly distributed, we have

$$\frac{1}{N} \left| \sum_{r=1}^{N} (A_r B_r + A_r B'_r + A'_r B_r - A'_r B'_r) \right| \leq 2 \tag{4.7}$$

or

$$|P(A, B) + P(A, B') + P(A', B) - P(A', B)| \leq 2, \tag{4.8}$$

where

$$P(A, B) = \lim_{N \to \infty} \frac{1}{N} \sum_{r=1}^{N} A_r B_r \tag{4.9}$$

Let us use our measuring apparatuses at spacetime events separated by a spacelike interval. As A and B are independent we may have $A = 1$ in the first apparatus and $B = 1$ in the second apparatus. We say that there is *locality* if a *rotation* of the second apparatus in order that it measures B' instead of B, does not influence the results obtained with the first apparatus. Under these conditions it is clear that Eq. (4.8) should be valid.

It is easy to see that the formalism of quantum theory leads to a breakdown of inequality (4.8) in some cases. Let us consider a two photon system emitted in a $J = 0 \longrightarrow J = 1 \longrightarrow J = 0$ transition of an atom, called a *SPS cascade*. When the photons are detected in opposite directions (we will take the $+z$ and $-z$ directions) they have the same polarization, since the excited state of the atom is spherically symmetric and its decay is due to electromagnetic interaction which is also rotationally invariant. The final state of the atom plus photons is then spherically symmetric and *entangled*, the various states of the atom being correlated with the states of the photons. We can partially lift the entanglement using collimators which select the pairs of photons moving in the $+z$ and $-z$ directions. Then the resulting state for the photon pair ends in this case only with rotation symmetry along the z-axis.

Let $\mathcal{H}^p$ be the spin space for one photon[20], and let $|\chi_x\rangle$ and $|\chi_y\rangle$ be respectively the states of the photon with linear polarization along the x- and y- axes. As is well known, a photon state with linear polarization forming an angle a with the x- axis in the xy-plane is described by

$$|\chi_a\rangle = \cos a |\chi_x\rangle + \sin a |\chi_y\rangle. \tag{4.10}$$

The orthogonal state is

$$|\chi_a^\perp\rangle = -\sin a |\chi_x\rangle + \cos a |\chi_y\rangle. \tag{4.11}$$

An apparatus designed to test the linear polarization of a photon polarized at an angle a relative to the x-axis is represented by the projector

$$\mathbf{P}_a = |\chi_a\rangle\langle\chi_a|. \tag{4.12}$$

It is convenient to introduce the operator

$$\begin{aligned}\mathbf{A}_a &= 2\mathbf{P}_a - 1 \\ &= \cos 2a (|\chi_x\rangle\langle\chi_x| - |\chi_y\rangle\langle\chi_y|) + \sin 2a (|\chi_x\rangle\langle\chi_y| + |\chi_y\rangle\langle\chi_x|).\end{aligned} \tag{4.13}$$

The observable corresponding to the operator given by Eq. (4.13) will be denoted by A_a and will be called *spin projection*, since its eigenvalues ± 1 correspond to the eigenvalues 1 and 0 of $\mathbf{P}_a$.

For a photon pair the Hilbert (spin) space is $\mathcal{H}_1^p \otimes \mathcal{H}_2^p$, with $\mathcal{H}_1^p \approx \mathcal{H}_2^p \approx \mathcal{H}^p$. Only the state vectors $|\Psi_+\rangle, |\Psi_-\rangle \in \mathcal{H}_1^p \otimes \mathcal{H}_2^p$, where

$$|\Psi_+\rangle = \frac{1}{\sqrt{2}}(|\chi_x^1\rangle \otimes |\chi_x^2\rangle + |\chi_y^1\rangle \otimes |\chi_y^2\rangle) \tag{4.14}$$

$$|\Psi_-\rangle = \frac{1}{\sqrt{2}}(|\chi_x^1\rangle \otimes |\chi_y^2\rangle - |\chi_y^1\rangle \otimes |\chi_x^2\rangle) \tag{4.15}$$

are invariant under rotations around the z-axis and since only $|\Psi_+\rangle$ is even under reflection, it must be the one describing our photon pair coming from the *SPS cascade*.

Let us take A and A' as spin projections for the first photon, having an angle $a(a')$ with the x-axis in the xy plane, and B and B' as spin projections for the second photon, having an angle $b(b')$ with the x-axis in the xy plane. The spin projections operators $\mathbf{A}_a, \ldots \mathbf{B}_{b'}$ are all given by equations similar to (4.13). We can immediately verify that for our photon pair the product $\mathbf{A}_a \otimes \mathbf{B}_b$ has eigenvalues ± 1, which correspond respectively to identical and opposite results of the two measuring apparatuses. The average value of the observable $\mathbf{A}_a \otimes \mathbf{B}_b$ is the so-called *correlation function* of the outcomes of the two measuring devices. We have

$$Q(\mathbf{A}_a \otimes \mathbf{B}_b) \equiv \langle \mathbf{A}_a \otimes \mathbf{B}_b \rangle \equiv \langle \Psi | \mathbf{A}_a \otimes \mathbf{B}_b | \Psi \rangle = \cos 2(a-b) \quad (4.16)$$

Taking now $2a = 0°$, $2b = 135°$, $2a' = 45°$, $2b' = 90°$, and writing in the first member of Eq. (4.8) our Q's in place of the P's we get

$$1 + \frac{\sqrt{2}}{2} + \frac{\sqrt{2}}{2} > 2 \quad (4.17)$$

showing a *breakdown* of Eq. (4.8).

This simple example shows very clearly some kind of *nonlocality* in quantum physics, because physical results in it cannot be explained if there only exists information about local properties of quantum objects in spacetime. The *nonlocality* means that somehow one particle influences the other even when they are separated by a space-like interval.

4.4 BELL'S PROOF

Let us now reproduce Bell's original proof of his inequalities[1], showing that the validity of both quantum theory and special relativity is *inconsistent* with any relativistic probabilistic process. Another wording for this is the impossibility to reproduce quantum mechanical results by any *theory of local hidden variables*.

Suppose that at a certain spacetime point e_1, the value of A_a is measured and at the spacetime point e_2, separated from e_1 by a space-like interval B_b is also measured. Both observables can take the values ± 1, the indices a, b meaning as before, dependence of these values on the chosen directions.

Suppose now that the result of a measurement A depends on both the direction a and on some *hidden parameter* λ. This parameter is *hidden* to the observer at the time when A is measured, but can in principle be known as some condition in the general past of the particles occupying the spacetime points e_1 and e_2. The result of a measurement of B also depends on the direction b and on the same parameter λ. This parameter, by hypothesis, was defined somewhere in the Ω-intersection of the light cones of the past for the spacetime points e_1 and e_2.

Locality of the hidden variable means that A does not depend on b and B does not depend on a. If this is the case, any correlations for A and B can be conditioned only by the general past in which λ is defined.

Then, if $\rho(\lambda)$ is the probability distribution of the parameter λ in Ω, we can write the mathematical expectation value of the product $A_a B_b$ as

$$P(A_a B_b) = \int_\Omega \rho(\lambda) \overline{A}(a, \lambda) \overline{B}(b, \lambda) d\lambda \quad (4.18)$$

where $\overline{A}(a, \lambda), \overline{B}(b, \lambda)$ are the averaged values of A_a and B_b if during the measurements there is some dependence of the obtained values on the apparatus measuring these values (the so called *context-dependent* hidden variables, see Chapters 9, 10 and 11). In any case $|\overline{A}| \leq 1, |\overline{B}| \leq 1$.

Let a' and b' be alternatives for the positions of the measuring apparatuses. Then,

$$P(A_a B_b) - P(A_a B_{b'})$$
$$= \int_\Omega \rho(\lambda) \left[\overline{A}(a,\lambda)\overline{B}(b,\lambda) - \overline{A}(a,\lambda)\overline{B}(b',\lambda) \right] d\lambda$$
$$= \int_\Omega \rho(\lambda) \left\{ \{\overline{A}(a,\lambda)\overline{B}(b,\lambda)[1 \pm \overline{A}(a',\lambda)\overline{B}(b',\lambda)]\} \right\} d\lambda$$
$$- \int_\Omega \rho(\lambda) \left\{ \{\overline{A}(a,\lambda)\overline{B}(b',\lambda)[1 \pm \overline{A}(a',\lambda)\overline{B}(b,\lambda)]\} \right\} d\lambda \quad (4.19)$$

From $|\overline{A}| \leq 1, |\overline{B}| \leq 1$ it follows that

$$|P(A_a B_b) - P(A_a B_{b'})|$$
$$\leq \int_\Omega \rho(\lambda) \left\{ [1 \pm \overline{A}(a',\lambda)\overline{B}(b',\lambda)] - [1 \pm \overline{A}(a',\lambda)\overline{B}(b,\lambda)] \right\} d\lambda \quad (4.20)$$

or

$$|P(A_a B_b) - P(A_a B_{b'})| \leq 2 + P(A_{a'} B_{b'}) + P(A_a B_b) \quad (4.21)$$

$$|P(A_a B_b) - P(A_a B_{b'})| + |P(A_{a'} B_{b'}) + P(A_a B_b)| \leq 2 \quad (4.22)$$

These are *Bell's inequalities*.

It is easy to see that in quantum theory, where $P(A_a B_b)$ is given by $Q(\mathbf{A}_a \otimes \mathbf{B}_b)$, etc..., that these inequalities are violated.

Following Clauser, Horne et al[3], another way of looking at Bell's inequalities is to consider situations for which the usual probability theory is valid and there are statistical correlations. If $\wedge$ represents the classical conjunction of probability calculus, and then if $P(a \wedge b) \neq P(a)P(b)$ and the correlations are conditioned only by the general past, denoting these conditions by λ we obtain

$$P_{12}(a \wedge b|\lambda) = P_1(a|\lambda) P_2(b|\lambda), \quad (4.23)$$

$$P_{12}(a \wedge b) = \int d\lambda \rho(\lambda) P_1(a|\lambda) P_2(b|\lambda). \quad (4.24)$$

Denoting now $P_1(a|\lambda) = A$, $P_2(b|\lambda) = B$, $P_1(a'|\lambda) = A'$, $P_2(b'|\lambda) = B'$ and calling

$$\kappa(\lambda) = AB - AB' + A'B + A'B' - A' - B \quad (4.25)$$

it follows from the inequalities $0 \leq A, A', B, B' \leq 1$ the so called Clauser-Horne inequalities:

$$-1 \leq \kappa(\lambda) \leq 0. \quad (4.26)$$

If we do not know the value of λ and average over its probabilistic distribution, we arrive at Bell's inequalities. From this point of view the breakdown of Bell's inequalities means the invalidity of Eq. (4.26) and thus, of the standard probability theory for quantum objects.

Notes

1. There are some scientists that still have doubts concerning the results of Aspect's experiments, see, e.g.,[12,13].

2. *EPR* discussed a situation involving combinations of coordinates and momenta.

3. One of the authors (W. A. R.) would like to comment that recently it has been found[14-16] that all relativistic wave equations have solutions with arbitrary speeds $0 \leq v/c < \infty$. In particular, there exist superluminal solutions of Maxwell equations in vacuum. Computer simulations produced in[16] showed that these waves can be generated in our physical world and this seems to be confirmed by a recent experiment[17] (see also [18] for comments on [17]).

4. According to Bell[19]: "In particular we will exclude the notion of 'observable' in favor of that of '*beable*'. The beables of the theory are those elements which might correspond to elements of reality, to things which exist. Their existence does not depend on 'observation'. Indeed observation and observers must be made of beables."

References

[1] J. Bell, On the Einstein-Podolski-Rosen paradox, *Physics* **1**, 195–200 (1965).

[2] J. F. Clauser and A. Shimony, Bell's theorem—experimental tests and implications, *Rep. Prog. Phys.* **41**(12), 1881–1927 (1978)

[3] J. F. Clauser, M. A. Horne, A. Shimony and R. A. Holt, Proposed experiment to test local hidden-variable theories, *Phys. Rev. Lett.* **23**(15), 880–884 (1969).

[4] D. M. Greenberger, M. A. Horne, A. Shimony and A. Zeilinger, Bell theorem without inequalities, *Am. J. Phys.* **58**(12), 1131–1143 (1990)

[5] L. Hardy, Quantum mechanics, local realistic theories and Lorentz invariant realistic theories, *Phys. Rev. Lett.* **68** (20) 2981- 2984 (1992).

[6] A. Aspect, P. Grangier and G. Roger, Experimental tests of realistic local theories via Bell theorem, *Phys. Rev. Lett.* **47**(7), 460–463 (1981).

[7] A. Aspect, I. Dalibard and G. Roger, Experimental test of Bell inequalities using time-varying analysis, *Phys. Rev. Lett.* **49** (25) 1804–1807 (1982).

[8] B. d'Espagnat, The quantum theory and reality, *Scient. Am.* **241**(5), 128–140 (1979).

[9] H. P. Stapp, *Mind, Matter, and Quantum Mechanics* (Springer-Verlag, Berlin, 1993).

[10] A. Einstein, B. Podolski and N. Rosen, Can quantum mechanical description of physical reality be considered complete?, *Phys. Rev.* **47**, 777–780 (1935).

[11] D. Bohm and Y. Aharanov, Discussion of experimental proof of the paradox of Einstein, Rosen an Podolsky, *Phys. Rev.* **108**, 1070–1076 (1957).

[12] F. Selleri, *Quantum Paradoxes and Physical Reality* (Kluwer Academic Publishers, Dordrecht, 1990).

[13] J. P. Wesley, *Classical Quantum Theory* (Benjamin Wesley Publisher, Blumberg, 1996), p. 143.

[14] W. A. Rodrigues, Jr. and J. Vaz, Jr., Subluminal and Superluminal Solutions in Vacuum of the Maxwell Equations and the Massless Dirac Equation, in: *Proc. of the Conference: The Theory of the Electron* (Mexico-1995), edited by J. Keller and Z. Oziewicz, *Adv. Appl. Clifford Algebras* **7**(S), 457–466 (1997).

[15] W .A. Rodrigues, Jr. and J. E. Maiorino, A unified theory for construction of arbitrary speeds ($0 \leq v < \infty$) of the relativistic wave equations, *Random Oper. and Stoch. Equ.* **4**(4), 355–400 (1996).

[16] W. A. Rodrigues, Jr. and J. Y. Lu, On the existence of undistorted progressive waves (UPWs) of arbitrary speeds $0 \leq v < \infty$ in nature, *Found. Phys.* **27**(3), 435–508, (1997).

[17] P. Saari and K. Reivelt, Evidence of X-shaped propagation-invariant localized light waves, *Phys. Rev. Lett.* **74**(21), 4135–4138.

[18] E. Capelas de Oliveira and W. A. Rodrigues, Jr., Superluminal electromagnetic waves in free space, *Ann. der Physik* **7**(7-8), 654–659 (1998).

[19] J. S. Bell, *Speakable and Unspeakable in Quantum Mechanics* (Cambridge University Press, Cambridge, 1987), p. 174.

[20] A. Peres, *Quantum Theory: Concepts and Methods* (Kluwer Academic Publishers, Dordrecht, 1993).

5
COPENHAGEN INTERPRETATION

In a discussion with Einstein on the *EPR* issue, Bohr[1], who is considered the "father" of the Copenhagen interpretation (*CI*), made claims against Einstein's idea of *objective elements of reality* existing in the microworld. He believed that the description given by quantum theory is complete. Also he insisted on four important features of quantum physics, which are:

(a) "Impossibility of any acute separation of the behavior of the atomic objects and the interaction with the measuring devices which serve to define the conditions under which the phenomena appear."

(b) "The formalism does not allow pictorial representations, on customary lines, but aims directly at establishing relations between observations, obtained under well defined conditions."

(c) "It is imperative to realize that in every account of physical experience we must describe both experimental conditions and observations by the same means of communication as the one used in classical physics."

(d) "We have been forced... to reckon with a free choice on the part of nature between several possibilities."

The above four points of Bohr will be important for our investigation of the *CI*, because we must agree with von Weizsäcker[2] that "the interpretation in my view, has never been fully clarified. It needs an interpretation, and that will be its only defense."[1]

Here we shall present two *interpretations* of the *CI*. In Chapters 6 and 7, we shall discuss the *many worlds* interpretation and the *quantum logic* interpretation, which could also be considered as interpretations of the *CI*. In each case we present: (i) the main points of the given interpretation according to the view of its originator(s); (ii) the meaning of nonlocality in the interpretation; (iii) the criticisms that can be and/or have been raised against the interpretation.

A *common* point of all these interpretations is that the mathematical formalism of quantum theory provides a *complete description* of the phenomena occurring in the microworld.

Accepting this statement as *true*, then due to the nonexistence of a state vector which is a simultaneous eigenstate of noncommuting operators, and also due to the breakdown of Bell's inequalities, as, e.g., in the d'Espagnat version[6], we have the following conclusions:

The properties of *micro-objects,* as position, momentum, spin projections, particle number and local current density (in quantum field theory) do not exist as *elements of reality* or *beables*[7] as Bell likes to say. They exist (or *arise* as definite ones) only if they are measured.

This means, e.g., that only with respect to a set of measuring apparatuses that permit the measurements of a maximal set of commuting operators which specify, let us say, the *electron properties*, is that by measuring the particle number we can say: *one electron exists.* Relative to an apparatus that measures the electron/positron current density[8], there is no electron, but there is the local quantum electron/positron field.

If we have some apparatus that permits the measurement of the spin projection of the photon (polarization) on a given axis, then a definite value of this projection arises due to the measurement. A photon *as it is* cannot have any definite value for any of its spin projections. Even more, there is no such *object* as a *photon* without an apparatus preparing it as a *photon*.

Now, what is *nonlocality* from the point of view of the formalism of quantum theory?

Consider the *EPR* situation in an experiment where, e.g., a two-photon system, coming from an a cascade emission of an atom is prepared, in a special state which is an eigenstate of the permutation operator **P**, which commutes with the Hamiltonian of the atom. Since **P** does not commute with the operators of position or spin projections of individual photons, then if some *local* observable is measured for one of the photons, the wave packet collapse takes place and the state vector of the whole system *leaps* to a new state vector

This leap, as we know (Chapters 2 and 3), is not described by the Schrödinger equation and happens instantaneously. So, due to a measurement realized in particle number 1, particle 2 has its state changed and we can see manifestations of this change in experiments which show the breakdown of Bell's inequalities.

In all experiments showing *nonlocality* in quantum theory, it is important that we have the system prepared in such a way that its state vector is not an eigenstate of local operators. We do not have manifestations of nonlocality for situations where the state vector is the product of the state vectors of the individual particles.

Now, let us discuss some interpretations of all this.

5.1 ABSENCE OF QUANTUM OBJECTS

If quantum properties and quantum objects exist only if there are corresponding classical measuring apparatuses, we can say together with Ludwig[9] ("Do atoms exist?") that "only classical objects really exist: microparticles or quantum objects are only *relations* between classical macro-objects." This is the *operationalistic* philosophy in action in the domain of quantum theory. In order to understand this point of view of the unreality of quantum objects let us specify three types of *relations* between macro-objects[2]:

(a) Relations that can be described by some *carrier*. This carrier can be called an object and moves along some worldline in spacetime. For example, there is a relation between the hand of a boy, throwing a stone, and a broken window. This relation has as its carrier the stone, moving along a definite worldline in spacetime. Relations of this kind are described by classical mechanics, and also by classical statistical mechanics if somehow, there is ignorance of either the initial conditions or the laws of motion.

(b) Relations without a carrier moving along some definite worldline in spacetime, but which are described by some deterministic equation, representing the possibility that the observer has to define forces acting on probe objects in any point in spacetime. Usually, this kind of relation is the one between a source and the measuring apparatus (excluding here the primordial radiation in cosmology for which, according to a *majority view*, there is no material source). These relations are called *fields* (electromagnetic and gravitational).

(c) Relations between macro-objects without any (identifiable) carrier, neither a classical particle nor a field.

The first example of this kind is the so called Lorentz contraction of a macroscopic object in motion and the time dilation of a clock in motion in relativity theory. According to the orthodox interpretation of this theory these phenomena do not have a physical cause, being "effects" of the Lorentz geometry of spacetime.[3]

The second example is given by *quantum objects*. If, in the region of spacetime between the preparing and measuring apparatuses used to prepare and detect a quantum object, we put some particle as a *probe* body, than differently from the case where there is a field in that spacetime region, we will not register at any point any force, but it is probable that at some other point there will be some kind of action! That is why, differently from a field we usually register something called a *particle*, which manifests itself at the spacetime point where it is detected.

This micro-relation can be characterized by some parameters such as, e.g., mass, spin, charge. For their observation we must use apparatuses where Planck's (action) constant cannot be considered negligible. An electron, from this point of view is just a *relation* between complex macroscopic objects— the electronic pump and operations with it— and the photoemulsion where the *spot* appears. Some operations on the electronic pump are called *preparation* of the wave function of the electron. The appearance of the spot is called measurement. The individual appearance of a spot is unpredictable, in the sense that we cannot predict with certainty the electron's position. But anyway it is determined by the electronic pump, because without it there is no relation. This relation is consistent with special relativity in the sense that until now, in all experiments done, the random effects show up only after some time, and the ratio between the distance between the pump and the spot in the emulsion and the elapsed time is less than c, the velocity of light in vacuum. If we perform the operation called "preparation" of the electron in the initial state $|\Psi(t_0)\rangle$ many times, and use the Schrödinger equation to calculate $|\Psi(t)\rangle$, $t > t_0$, then we can effectively predict the *frequency* of the distribution of spots in the emulsion, and here (i.e., in the prediction of the frequencies), it is important to emphasize, we have strict determinism.

Protons, mesons, etc. are also names for more complex technical objects and manipulations with them by the use of devices such as accelerators, reactors, bubble chambers, etc. They are also relations between astrophysical objects like the sun, stars, pulsars and our measuring apparatuses on the earth.

From this point of view, different *names* for quantum objects are connected with different macroscopic invariants, which are the same regardless of the preparation and measurement apparatuses used. In common these apparatuses have either a strong or a high frequency electromagnetic field, and a high or low temperature, etc., when such parameters as mass, charge, spin and surely Planck's constant are not negligible. So, e.g., if we detect with an apparatus that at some point a given mass and charge become smaller and if we detect through another apparatus that in another place they become large, we can say that there is a propagation of a charged particle.

The wave function, from this point of view does not describe some existing object, but the preparing or measuring apparatuses. This is the reason why Bohr[1] did not like such words as *"the properties of the quantum object arise as definite in a measurement,"* because there is no object at all, and because if we use another measuring apparatus, another property will correspond to it. There is no *leap* due to the collapse of the wave function as something *objective* occurring with the quantum object. Changes of the wave function due to the Schrödinger equation or by wave packet collapse are only mathematical rules for calculating the probabilities for the values of observables as measured by appropriate measuring devices.

5.1.1 NONLOCALITY

Now, let us make some remarks on the question of nonlocality in this picture. We said before, that quantum theory respects relativity in the sense that its mathematical rules are such that the wave functions (for the one particle states) of the theory, that correspond to the different kinds of "particles", satisfy one of the so-called relativistic wave equations such as, e.g., the Klein-Gordon-Fock, Maxwell, Weyl or the Dirac equations, and until now we have not seen (with certain[4]) in any experiment which has been done so far a noncausal propagation, i.e., one in which the group velocity of the wave is greater than the velocity of light in vacuum. Despite this fact there is a problem with the *EPR* situation, as discussed in Chapter 2, e.g., in the Stapp's version of the breakdown of Bell's inequalities. If we have a non zero state vector for a two-particle system, which is an eigenstate of the permutation operator, then measurements at the point where one of the particles is found influences the results of measurements of the other particle at the same instant, whatever the distance between the particles. Nevertheless, such a *relation* exists only for individual results, not for the frequencies, this being the reason why it is impossible to use the *EPR* nonlocality effect to send a *signal* between two points separated by a space-like interval[5]. The relation is *hidden* in quantum indeterminism. To see non local correlations, we must compare results obtained at the two space-like separated points. Without such a comparison, it is impossible to see anything. In a sense this is similar to the Lorentz contraction, which cannot be seen by the observers which are at rest relative to the "contracted rod" but can be seen only when we compare the lengths of two rods (with the same rest length) as determined in a given inertial reference frame, where one of rods is at at rest and the other is in motion[6].

Recently, this nonexistence of the quantum object interpretation has been expressed in a somewhat new way by the son of N. Bohr, A. Bohr and by Ulfbeck[15]. These authors intend to have produced a *geometrical* interpretation of quantum theory. They said that according to Einstein, the gravitational field is explained as spacetime curvature, but *quantum objects* do not exist, there exists only empty Minkowski spacetime. This spacetime has the maximum number of Killing vectors in four dimensions, which implies in symmetries described by the Poincaré group (*PG*). The unitary representations of the *PG*, first obtained by Wigner[16], *are* the quantum objects. So, the symmetry is not to be understood as symmetry of a quantum object, but *symmetry itself is the object*. Energy, momentum, etc. are just the Lie algebra generators of the *PG* leading to observables. The Heisenberg uncertainty relations are, as is well known, consequences of the noncommutativity of translations and Lorentz transformations in the non relativistic limit. Pauli's exclusion principle appears as a consequence of identical operations performed by the experimenter. So, according to Bohr and Ulfbeck quantum theory is a consequence of relativity theory.

The difference between quantum and classical physics is claimed to be the difference between *sharp* (*exact*) operations when we have irreducible representations of the *PG*, and *unsharp* (*fuzzy*) operations, when we neglect Planck's constant and do not check the non Abelian nature of the *PG*.

Thus, this proposal (if we also take into account our previous discussion) permits us to say that if we perform a sharp operation with a macroscopic object we see a manifestation of quantum mechanics. If we perform an unsharp operation we see a manifestation of classical physics.

5.1.2 CRITICISMS

After this brief exposition of this *interpretation* of the *CI*, let us present some criticisms. But first, we must say that if we try to find some popular illustration of this picture for laymen, maybe the best example is that of someone superstitious, who says that he makes a relation between a black cat that crosses his path and some unpleasant event. He also claims that there is no *carrier* for the influence coming from the cat and, let us say, the letter that he receives informing that he has just lost his job.

Now, for observers who do not share the superstitions of this fictitious person, there is no correlation whatsoever between the events mentioned above, and this is the only (but surely serious) difference with respect to quantum particles. Three serious arguments against that picture are the following:

(i) According to the intuition of the majority of human beings, what seems to be the *truth* is that macrobodies are made of quantum particles and not the opposite statement that a quantum particle is some description of the behavior of macro-objects.

This intuition seems to be proved by the successes of many-particle quantum theory, showing that many thermodynamical properties of macrobodies can be understood on the basis of quantum theory. In this case, we do not use the preparation and measuring technique.

As we said before, quantum theory is more fundamental than any classical physical theory. This means that strictly speaking there are no objects whose behavior satisfies the classical theories and this means that there are no strictly *unsharp* operations. Only when we have a realization of actual infinity or some new fundamental constant (like, e.g., Planck's mass $m_P = 10^{-5}$ g in Penrose's theory[20]) is that we can *always* have classical objects. If Schrödinger's interfering cats can indeed exist, all *classical foundation* will be ruled out.

Also, modern Big Bang cosmology tells us that our classical world evolved from the quantum one, and this may then be presented as one of the reasons why quantum theory is more fundamental than classical theories.

(ii) In the S-matrix formulation of elementary particle theory, we use with success the notions of *intermediate states*, *virtual particles*, etc., which have many properties, like the *real* particles. If we follow the operationalist philosophy, we should call the virtual particles *relations between relations*. Surely, this would be very bad language and the standard *objective particle idea* seems more pragmatic.

(iii) There are situations in which particles can be measured without any classical preparation and then there is no relation between macro-objects. A very significative example is given by the photons of the primordial radiation that (according to the cosmologist majority view) has no source. This can be compared with Unruh radiation[21], which refers to the creation of particles from the vacuum as detected by an accelerated detector. Photons of this radiation, described by a thermal distribution, arise due to a special measurement by the accelerated detector of an operator (Lorentz boost) that does not commute with the field Hamiltonian which is valid in the inertial reference frame. If we are to follow A. Bohr, here we could say that we make two noncommuting sharp measurements; the first measurement prepares the Minkowski vacuum and the second one prepares the Rindler vacuum (see [22]). Anyway, in the case of the primordial radiation we do not have the possibility of using this wording.

As to A. Bohr's ideas, it is also important to say that quantum particles have many other observable properties (*internal degrees of freedom*) besides those associated with the generators and the Casimir invariants of the *PG*. We might say then contrary to his ideas, that Minkowski spacetime has its existence and its symmetries because there exist quantum objects, and that some of their properties fortunately correspond to generators of the *PG*. This reasoning would lead to the idea that the Minkowski spacetime is secondary to the quantum world and not the opposite.

Finkelstein's expression of *nonobjective physics* in his recent book[23] is closely related to A. Bohr's idea. According to Finkelstein, there are no quantum *objects*, but there are *acts*, and quantum theory is the theory of these acts. The world is constructed by acts and not from objects. To acts correspond some operations and they are described by operators. However, differently from A. Bohr, Finkelstein does not think that these operations are only the symmetry operations of the Poincaré group. According to him the mathematical structure of these operations is either a non distributive lattice or a semi-group.

Macroscopic objects as well as micro-objects also do not exist as independents objects of some external reality and are nothing more than acts described by an appropriate set of commuting operators. In particular, the possible approximate deterministic description of macro-objects is what permit us to identify *acts* with *facts* or *events*. On the other hand micro-objects are pure *acts*.

We can ask: what are these acts? Are these acts only the preparing and the measuring acts of the observer?

Finkelstein did not give a clear answer to these questions, but we guess that it could be: these are acts of nobody! Like in the philosophy of existentialism or in Buddhist thought, *acts* are primary to *facts*. The *existence* formed by free choices of acts is primary to *essence* constructed by it. The I itself is then just a name for some collection of acts and as Buddhists think, observers really do not exist.

In resume, despite the fact that Finkelstein's interpretation is well adapted to the mathematical formalism of quantum theory, here we have the same problem that already occurred in relation to A. Bohr's interpretation, namely: why the manifested world of events is mathematically associated with real numbers and not with noncommuting operators associated with observables? Also, why any observer does not see the operator but only one of its eigenvalues.

Whatever the answer, we see that going from acts to facts needs some extra intervention, and here we recognize once again our old problem of the role of the observer in quantum physics.

Continuing with the assumption that quantum theory is a complete description of the microworld, we present now a second interpretation of the *CI* that is derived from the works of von Neumann[24], Heisenberg[25], Fock[26] and Stapp[27].

5.2 OBJECTIVELY EXISTING POTENTIALITIES WORLD

This interpretation says that micro-objects exist and have their own reality, but this reality is different from the one we perceive in the macroscopic world and which we describe with classical physics. It is some abstract reality of *objectively existing potentialities* as Heisenberg and Fock liked to describe it. The word 'potentialities' means that only relative to a measuring apparatus and the observer behind it does a possible property of a quantum acquire a *yes/no* status and become an *actual event*. However, different from the potentialities of classical physics that exist only in our mind due to our ignorance of the real situation, the quantum potentialities objectively exist. A quantum object from this point of view is just some abstract set of observables. This set can close some algebra, e.g., it can be the Lie algebra of the *PG*. Then, similarly as in the A. Bohr's interpretation, we can identify *different* particles with a different irreducible unitary representation of the *PG*, and it is important to emphasize here that *objectively*, only the structure of this algebra and its representations exist. Concrete values of the energy, momentum or coordinates of a quantum particle arise as eigenvalues of the corresponding operators in the measurement process. The wave function describes the state of the quantum system and permits the prediction of the relative probabilities for the results of measurements of the observables of the system according to Born's frequency interpretation.

Recalling that the wave function can change deterministically according to the Schrödinger equation or due to wave packet collapse, we must say that according to von Neumann[24], London and Bauer[28] and Wigner[29], the wave packet collapse occurs due to a special role of the consciousness of the observer. The argument behind this statement is that it is because consciousness is not described by the Schrödinger equation that the possibility of a new kind of changement of the state vector (wave packet collapse) can arise.

Following von Neumann, we can say that in any actual measurement, we have a chain: (i) quantum particle; (ii) measuring apparatus, also composed of quantum particles; (iii) observer with his perception organs. The observer gets some information about the particle due to the measurement process. It is always possible to consider the perception organs of the observer, eyes, ears, etc., as other measuring apparatuses, which measure properties of the initial measuring apparatus (which interacted with the quantum particle). These organs are also composed of quantum particles, so that the whole system composed of the particle, the measuring apparatus, perception organs, can be described by the Schrödinger evolution.

But in the end, as observed by von Neumann, there is some *ultimate observer*, who is the *subject* of a cognitive relation and who gets information about the *object*. This *abstract ego* is not described by any Schrödinger equation and it is he (or she) who produces the wave packet collapse, which implies in definite positions for pointers, etc., as discussed in Chapter 3. Such an *effect*, which can only be described by a non unitary transformation, implies then, in a breakdown of the Schrödinger equation.

An important role, according to von Neumann is played by the *principle of displacement of the frontier subject-object,* which says that the wave packet collapse cannot depend on where we put a frontier between the measured object and the measuring device. So, it is possible to put this frontier between our microparticle and the first measuring device, which is then treated classically, or between our eyes and the particle, plus the first measuring device that in this case must be treated quantum mechanically. We can go even further, placing the frontier inside our brain...

What is really fantastic is that this frontier can be moved not only in space, but also in time. So, considering the measuring device as a continuation of our perception organs (see Fock[26]), we can say the following: some apparatus made a definite check *before* I, as a conscious subject register the event, but it is possible to place this frontier only at this instant *now* inside my brain, when the collapse occurs due to its registration by my consciousness!

London and Bauer developed von Neumann's idea, in their book[28] and tried to connect it with some special property of consciousness called introspection. Similar ideas were discussed by Pauli[30] in his talks with Jung about the difference between consciousness and subconsciousness and recently Stapp[19,27] also considered the idea of awareness.

According to London and Bauer[28], a complex system, quantum object $\mathcal{X}$, apparatus $\mathcal{Y}$ and observer $\mathcal{Z}$, can be described by the state vector,

$$|\Psi(\mathcal{X}, \mathcal{Y}, \mathcal{Z})\rangle = \sum_k a_k |u_k(\mathcal{X})\rangle \otimes |v_k(\mathcal{Y})\rangle \otimes |w_k(\mathcal{Z})\rangle \qquad (5.1)$$

where $|w_k(\mathcal{Z})\rangle$ are different states of the observer. Being part of the whole system, the observer is not described by some pure state and the density matrix for him is not a mixture of states. But then, the quoted authors say: "For him (observer) only the object $\mathcal{X}$ and the apparatus $\mathcal{Y}$ belong to the external world. On the contrary, with himself he has special relations: he has some well known capacity, which can be called capacity of self observation (introspection):without any agent he can make a check of his own state. Due to this self-knowledge, he can cut the chain of statistical connections expressed by the wave function $\sum_k a_k |u_k(\mathcal{X})\rangle \otimes |v_k(\mathcal{Y})\rangle \otimes |w_k(\mathcal{Z})\rangle$, saying: I am in the state $|w_k(\mathcal{Z})\rangle$ or I see "$g = g_k$" (g the position of the pointer), or directly: $f = f_n$ (f-property of the object). So, it is the *consciousness of the observer* that is responsible for the transformation of the density matrix into a mixture of pure states and that due to *introspection* produces a new wave function for the object, $|\psi'(\mathcal{X})\rangle = |u_l(\mathcal{X})\rangle$."

It is important that here consciousness plays an *active* role! In classical physics it only *passively* registers what really occurred in *objective* nature. Here it really *changes* or *organizes* (as C. de Beauregard likes to say[31−34]) the world. Even without looking at the pointer of the apparatus, the conscious observer will say that the apparatus *must* be in some mixture of states, i.e., with this or that probability it is in some pure state. But this mixture of states arises because he, the conscious observer, can find *himself* as *conscious* only in some definite pure state. The whole system, object plus apparatus plus observer changes in such a way that the wave function $|\Psi\rangle$ becomes the mixture described by the density matrix

$$\rho = \sum_k |a_k|^2 \mathbf{P}_{u_k v_k w_k} \qquad (5.2)$$

and then is fixed at some $|\psi_k\rangle = |u_k\rangle \otimes |v_k\rangle \otimes |w_k\rangle$.

If this interpretation is true, then the wave packet collapse can be described as the process of going from *subconsciousness* to *consciousness*, as *awareness* of something objective, and then physicists must be urged to consult psychologists and neurophysiologists for this possibility[7]!

5.2.1 NONLOCALITY

What is nonlocality in *EPR*-experiments from the point of view of this interpretation?

The wave function corresponds to some state of *knowledge* and has something psychological in it. But, it now describes some *objectively existing potentiality* which becomes reality if some measurement occurs. To see how this works to produce an explanation of the *EPR*-situation, let us consider two particles with spin 1/2 which are prepared in the singlet state, and then go into opposite directions. Consider the situation[35] when the spacetime coordinates of the correlated particles define a spacelike interval (Fig.(5.1)). Now, suppose that at the spacetime point e_1 we measure the spin projection S_z for the particle 1 and get $S_z = +1/2$. Observer 1 (at rest in the inertial reference frame I_1) using the rules of quantum theory will say : due to the wave packet collapse simultaneously (according to his time order) with the event e_1, particle 2 will *acquire* (or will be) in a quantum state such that $S_z = -1/2$.

We can then ask : is the acquisition by particle 2 of the value $S_z = -1/2$ an *event* in spacetime? It is easy to see that the answer is *negative*, if we accept the majority view interpretation of relativity theory which excludes the existence of a preferred frame which determines the time order univocally.

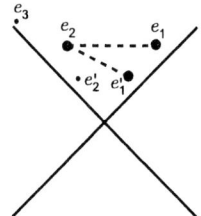

Figure 5.1. Is e_2 an event?

The *proof* of the statement (under the above condition) is as follows: suppose that it *is* an event. Then, we have two simultaneous events, the ones corresponding to the realizations of $S_z = 1/2$ for particle 1 and $S_z = -1/2$ for particle 2 according to the time order defined by observer 1 who took the measurement for particle 1. But, now let us look at the above situation from the point of view of an observer 2 at rest in the inertial reference frame I_2, which is in motion relative to I_1 and thus has a different time order. According to observer 2, the event simultaneous with event e_2 is e'_1, not e_1. So, e'_1 must be the event $S_z = +1/2$ for particle 1, which for observer 2 is previous to event e_1. The same reasoning can be used for e'_1 when e'_2 is simultaneous in some frame to e'_1, and so on. These arguments imply in the pre-existence of definite values of S_z for both particles, independently from any observation. But as we showed before, this will lead to the validity of Bell's inequalities. Since quantum theory implies in the breakdown of Bell's inequalities we arrive at the conclusion that $S = -1/2$ for particle 2 at e_2 is not an event if it is not observed.

In order to become an *event*, it is necessary for the second observer to measure at e_2 some observable, which can either be S_z or another which does not commute with it, such as, e.g., S_x or something else. The second observer can be substituted in this case by some measuring apparatus, but the first observer must have the possibility to look into his future on the register of the device.

We can ask: why, despite the appearance of event e_1 when the observer 1 got the definite wave function at e_1, in e_2 there is no event, but only an objectively existing possibility, despite the fact that the wave function of the correlated particles is definite for observer 1 after event e_1?

The answer is that a quantum object (here, the correlated pair) is the totality of its properties, described by noncommuting operators[8]. This totality is not described as an *object* in *spacetime* which is a set of events! But, when observer 1 makes his choice to measure one of the noncommuting operators at e_1, at point e_2 there is still the possibility of choosing what to measure there. It is only when a choice is made at e_2 by putting the measuring apparatus there that the event e_2 *arises*.

5.2.2 CRITICISMS

Let us now present two *criticisms* of the above interpretation.

5.2.2.1 The Wigner friend paradox. If it is the consciousness of the observer which is responsible for the wave packet collapse, imagine that a quantum system is observed by two observers, *Wigner and his friend*[37]. They both look at the position of the pointer of an apparatus which measures a given observable of the quantum system. According to the *subjective* interpretation, the obtention of an eigenstate of the operator occurs not in an *objective* way, independently of observers, but because due to introspection consciousness must find itself in some definite pure state. But then, why do different observers see the same state, leading to the same position of the pointer?

There are different attempts to solve this paradox. One of them is connected with von Neumann's idea, that the *absolute ego of the observer is only one*. There is always one *subject* of cognition which is *ME*, all other human beings for *ME* are objects. But, this *ME* is external to the objective universe and cannot be identified with any object. *ME* as John, Maria, Paul, etc, is objectivistic. So, this *Me* is the same in John, Maria, Paul, etc. and this is the reason why there is no Wigner's friend paradox.

Another solution is the *world consciousness idea*, that says that when I am conscious of something external, I am just participating in some nonlocal world consciousness, where introspection takes place (see Squires[38]).

Stapp[5] suggested another idea in which he proposes to consider *leaps* in quantum physics as objective actual events, occurring in nature, independently of any observer, but where our consciousness with its introspection is to be considered as a consequence or manifestation of these quantum events.

5.2.2.2 The problem of quantum processes before the existence of conscious observers. This problem does not exist, of course, for the adherents to von Neumann's idea, because when some observer says something about past eras he always discusses this something in terms of present information and then refers to dinosaurs, etc., as if he were present in that epoch.

Even more, following Wheeler[39], we can come to the idea of a *participatory universe*, where I, as observer today, participate in the quantum creation of the universe, which starting with the Big Bang, creates the particles, etc., and then leads to my actual body and apparatuses which I use to observe this creation. The past and the very notion of time with its *before* and *after* are defined by the observer now, in such a way that quantum processes in the earlier universe go, as if *oriented* to the production of the future observer, which by participation originates this universe. Due to the principle of movement of the frontier between object and subject in time, it becomes possible to say that decays of the uranium in Ocklo^9, millions of years ago before man appeared on the earth, happened the same way as now. The wave packet collapse could be moved from the now of modern observers to the past.

Following Stapp's hypothesis, we can speculate about the origination of human consciousness from some non human consciousness, due to the leaps that occurred before the origination of man.

Closely related to Stapp's idea is d'Espagnat's[35,40] that says that consciousness and physical (or empirical) reality should be considered as two complementary aspects of reality. This complementarity consists in the fact that each aspect contributes decisively to a greater characterization and to a greater specification of the other one. Nevertheless what separates this concept from idealism[10] is insistence on the difference between mind independent reality (which is holistic, beyond spacetime and hardly knowable by us, although our great mathematical laws may provide a glimpse of it) and empirical reality which is the set of phenomena in Kant's sense, that it is dependent on our mental and sensorial structures. Also d'Espagnat said that "empirical reality and consciousness timelessly generate one another, so to speak, within mind-independent reality. An electron, e.g., as a notion referring to the physical part of the existing whole, typically belongs to the empirical reality."

Another view concerning the relation between consciousness and physics can be attributed to Pauli[30] who said that our universe is a psychological universe in which there is unity between matter and consciousness. From this point of view, the distinction between objects, apparatuses and the observer means that any quantum object, like, e.g., an electron, is a nonlocal *physical and psychic unity*, such that the psychic property manifests itself in some other object—consciousness of the observer—as this or that property of the electron.

This idea, of course, implies that the universe cannot exist without observers, even if these observers can only be in the future of the quantum objects.

It seems that Wheeler and d'Espagnat's ideas can be considered as formulations of the *strong anthropic principle*[41]: our quantum universe necessarily must produce observers at some stage of its evolution.

In order to exhaustively describe the Copenhagen interpretation which considers that there is no need to change the mathematical theory behind quantum theory, we discuss in the next two chapters what we think can be considered as other interpretations of mathematical formalism: the many worlds interpretation of Everett-Wheeler-DeWitt and the quantum logic interpretation.

To conclude this long Chapter, let us observe that many physicists have the firm opinion that consciousness has nothing to do with quantum theory and that there must be some other way to deal with its paradoxes[20].

Notes

1. Of course, as noted by Peres[3], experts disagree on what is the meaning of the Copenhagen interpretation, e.g., Ballentine[4] says that the interpretation is the claim that "a pure state provides a complete and exhaustive description of a single system." Stapp[5] calls Ballentine's statement the *absolute-Ψ interpretation* and says that "critics often confuse the Copenhagen interpretation which is basically pragmatic, with the diametrically opposed absolute-Ψ interpretation."

2. At the long run, all these types of relations are nothing more than sensation relations, if we follow positivist philosophy.

3. W. A. R. would like to say that according to his point of view as expressed in[10], the non physical causation of the Lorentz contraction and time dilation phenomena was not Einstein's final view on these phenomena, as clearly illustrated in his writings[11].

4. We have already mentioned in footnote no. 3 of Chapter 4 that recently it has been discovered that all relativistic wave equations have solutions with arbitrary speeds $0 \leq v < \infty$. It is important also to say here that, contrary to common belief, there are solutions of the relativistic wave equations (describing quantum particles) such that the group velocity can be different from the particle velocity[12]. Another point worth having in mind is the recent experiments done by Nimtz[13] with the propagation of microwaves in wave guides that show superluminal propagation (see also [14]).

5. This statement is true only if there is no influence of the mind over matter as suggested by parapsychology experiments, e.g., by Schmidt[17] and Jahn and Dunne[18]. In this connection, see also the recent paper by Stapp[19], where he presents a modification of quantum theory that according to him permits the intelligibility of the parapsychology experiments described in[17,18]. Of course, if the parapsychological effects are true facts in the world we live in, then a breakdown of Lorentz invariance is one of the consequences...

6. One of the authors, A. A. G., is indebted to his teacher V. A. Fock for this comparison.

7. See, in this respect, the recent book by Eccles[36].

8. In Chapter 7, we also define a quantum object (system) as a non distributive lattice.

9. Ocklo is a place in Africa, where natural reactions of uranium decay took place (as in nuclear plants today) in a vast area in prehistoric time.

10. The text here follows private communication that d'Espagnat sent to the authors.

References

[1] N. Bohr, *Atomic Physics and Human Knowledge* (Science Editions, New York, 1961).

[2] C. F. von Weizsäcker, in: *Quantum Theory and Beyond: Essays and Discussions Arising from a Colloquium*, edited by E. Bastin (Cambridge University Press, Cambridge, 1971), pp.25–31..

[3] A. Peres, *Quantum Theory: Concepts and Methods* (Kluwer Academic Publishers, Dordrecht,1993).

[4] L. E. Ballentine, The statistical interpretation of quantum theory, *Rev. Mod. Phys.* **42**, 358–381 (1970).

[5] H. P. Stapp, The Copenhagen interpretation, *Am. J. Phys.* **40**(8), 1098–1116 (1972).

[6] B. d'Espagnat, Quantum theory and reality, *Sci. Am.* **241**(5), 128–140 (1979).

[7] J. S. Bell, *Speakable and Unspeakable in Quantum Mechanics* (Cambridge University Press, Cambridge, 1987).

[8] W. Heisenberg, in: *Niels Bohr and the Development of Physics*, edited by W. Pauli (McGraw-Hill, New York, 1955), pp. 12–29.

[9] G. Lüdwig, *An Axiomatic Basis for Quantum Mechanics*, vols.1 and 2 (Springer-Verlag, Berlin, 1985).

[10] W. A. Rodrigues, Jr. and M. A. F. Rosa, The meaning of time in relativity theory and Einstein's later view of the twin paradox, *Found. Phys.* **19**(6), 705–727 (1989).

[11] A. Einstein, in: *Albert Einstein: Philosopher-Scientist*, edited by P. A. Schilpp (Cambridge Univ. Press, London, 1970), pp.1–94.

[12] V. Barashenkov and W. A. Rodrigues, Jr., Launching of non-dispersive sub— and superluminal beams, *N. Cimento B* **113**(13), 319–328, 1998.

[13] G. Nimtz, in: *Proceedings of Adriatic Research Conference: Tunneling and its Implications* (07/30-08/02,1996), edited by D. Mugnai, A. Ranfagni, L. S. Schulman (World Scientific Publishing Co., Singapore, 1997), pp.223–237.

[14] S. Esposito, Classical $v_g \neq c$ solutions of Maxwell's equations and the photon tunneling effect, *Phys. Lett. A* **225** (4-6), 203–209 (1997).

[15] A. Bohr and O. Ulfbeck, Primary manifestation of symmetry-origin of quantal indeterminancy *Rev. Mod. Phys.* **67**(1), 1–35 (1995).

[16] E. Wigner, On the unitary representations of the Lorentz group, *Ann. Math.* **40**, 140–204 (1939).

[17] H. Schmidt, Collapse of the state-vector and the psychokinetic effect, *Found. Phys.* **12**(6), 565–581 (1982).

[18] R. G. Jahn and B. J. Dunne, *Margins of Reality: The Role of Consciousness in the Physical World* (Harcourt Brace Jovanovich Publ., San Diego, 1987).

[19] H. P. Stapp,, Theoretical model of a purported empirical violation of the predictions of quantum theory, *Phys. Rev. A* **50**(1), 18–22 (1994)

[20] R. Penrose, *Shadows of the Mind: A Search for the Missing Science of Consciousness* (Oxford University Press, Oxford, 1994).

[21] W. G. Unruh,, Notes on black hole evaporation, *Phys. Rev. D* **14**(4), 870–892 (1976).

[22] A. A. Grib, S. G. Mamayev and V. M. Mostepanenko, *Vacuum Quantum Effects in Strong Fields* (Friedmann Laboratory Publishing, St. Petersburg, 1994).

[23] D. Finkelstein, *Quantum Relativity* (Springer-Verlag, Berlin, 1996).

[24] J. von Neumann, *Mathematical Foundations of Quantum Theory* (Princeton University Press, Princeton, 1955).

[25] W. Heisenberg, *Physics and Philosophy* (Harper and Row, New York, 1958).

[26] V. Fock, *Quantum Physics and the Structure of Matter*, (in Russian) (LGU Press, Leningrad, 1965).

[27] H. P. Stapp, *Mind, Matter, and Quantum Mechanics* (Springer-Verlag, Berlin, 1993).

[28] F. London and F. Bauer, *La Théorie de l'Observation en Mécanique Quantique* (Hermann & Cie., Paris, 1939).

[29] E. P. Wigner, Die Messung quantenmechanischer operatoren, Z. *Phys.* **133**, 101–108 (1952).

[30] K. V. Laurikainen, *Beyond the Atom: The Philosophical Thought of Wolfgang Pauli* (Springer-Verlag, Heildelbeg, 1986).

[31] O. C. de Beauregard, Einstein-Podolski-Rosen non-separability and Feynman non-locality, *Phys. Lett. A* **60**(2), 93–95 (1977).

[32] O. C. de Beauregard, Time symmetry and the Einstein paradox, *N. Cimento B* **42**(1), 41–64 (1977).

[33] O. C. de Beauregard, Time symmetry and the Einstein paradox. 2, *N. Cimento B* **51**(2), 267–279 (1979).

[34] O. C. de Beauregard, CPT invariance and interpretation of quantum mechanics, *Found. Phys.* **10**(7-8), 513–530 (1980).

[35] B. d'Espagnat, *Conceptual Foundations of Quantum Mechanics* (W. A. Benjamin, Reading, MA, 1976).

[36] J. C. Eccles, *How the Self Controls its Brain* (Springer-Verlag, Berlin, 1994).

[37] E. P. Wigner, in: *The Scientist Speculates,* edited by I. J. Good (W. Heinemann, London, 1961), pp. 284–302.

[38] E. Squires, *Conscious Mind and the Physical World* (IOP, Bristol, New York, 1990).

[39] J. A. Wheeler, in: *Mathematical Foundations of Quantum Mechanics*, edited by A. R. Marlow (Academic Press, New York, 1975), pp. 9–48.

[40] B. d'Espagnat, *Veiled Reality* (Addison-Wesley Publ. Co., Reading, MA., 1995).

[41] J. Barrow and F. Tippler, *Anthropic Principle* (Oxford University Press, Oxford, 1986).

6
MANY WORLDS INTERPRETATION (MWI)

6.1 THE ROLE OF THE OBSERVER

This interpretation of quantum theory has been first proposed by Everett[1] and then discussed by Wheeler[2] and DeWitt[3]. Its claim is that it is a *realistic interpretation* of the wave function and it is popular among those scientists who try to discuss *quantum cosmology*.

Despite sharing the same mathematical formalism in common with the Copenhagen interpretation, the fact is that the *MWI* also has many different *interpretations*, and here we will follow Bell's[4].

Differently from the *objectively existing potentialities* idea, the authors of the *MWI* say that all possibilities encoded in the wave function take place, but in different worlds, in different universes! Explicitly, if we have the wave function which is a linear combination of the eigenvectors of some observable A represented by the Hermitian operator **A**,

$$|\Psi\rangle = c_1|1\rangle + c_2|2\rangle + ... + c_i|i\rangle + ... \tag{6.1}$$

and if some measurement is taken by an observer and as its result $|1\rangle$ occurs, then in another world $|2\rangle$ occurs, and in some other world $|3\rangle$ occurs, etc.

Because everything possible occurs in some world, every observer must have copies of himself in each one of the possible worlds, in such a way that each observer's copy will detect one possibility, the one that takes place for the copy of the world where (for unknown reasons) he perceives himself.

The *realism* of this interpretation consists in the claim that everything is described by the wave function and its Schrödinger evolution. So, the claim is that there is no collapse of the wave function in nature. Instead, there is a *process of splitting* of the universe into many universes, each time when some observer measures in a quantum system described by the superposition given by Eq. (6.1). In each one of the split worlds the possibility $|i\rangle$ occurs.

The process of splitting seems strange to almost all the readers of the original literature, because the authors say that everything is described by the Schrödinger equation and nothing else.

Surely, physical splitting, like doubling, etc., contradicts the conservation laws of charge, energy, etc. So, the only consistent idea seems to be:

(i) Since the beginning, there is an infinite number of copies of our universe and of its particles, and there, all the possibilities encoded in the wave functions of quantum systems occur.

(ii) Contrary to standard notions, there are *many* (indeed, infinite) pasts, instead of only one past. As Bell puts it[4], the authors of the *MWI* propose to make a more serious change in our idea of time, than Einstein did, with his proposed elimination of absolute simultaneity.

(iii) The actual infinite number of worlds evolve according to some master Schrödinger equation.

(iv) For the actual infinity of worlds with the same wave function, the frequency to observe some definite result in the measurement of observable A is given by Born's quantum mechanical rule (See section 2.1).

What is the role of the *observer* in this picture? There are at least three unusual features associated with the observer's concept, which are:

(a) an infinite number of copies of each observer exists in the different infinite number of universes.

(b) the observer has memory of his *past* experiences.

(c) the observer cannot be conscious of the existence of his copies, and imagines that he is the unique one.

So, the logical conclusion is that splitting occurs as a consequence of (b) and (c). At some moment of time t the observer checks some quantum system and finds that it must be described by some wave function $|\Psi\rangle$, given by Eq. (6.1). Then, at a later instant of time, the observer that cannot identify himself with the infinite number of his copies, by chance identifies himself with the copy in the world which evolved by the Schrödinger evolution from the other worlds (because there are many and not one past) to $|1\rangle$.

So, this observer with memory of $|\Psi\rangle$ will see $|1\rangle$. If he has this experience many times, i.e., if he follows the evolution of many quantum systems described by $|\Psi\rangle$, he will observe the frequency of obtaining $|1\rangle$ as predicted by quantum theory.

In the usual Copenhagen interpretation, different bases will be chosen by the observer, depending on which observable from the noncommuting set he decided to measure. But, DeWitt claims that in the *MWI*, observers are not different from machines, in the sense that nothing depends on their will. There is no free will for the observer here: a given apparatus or some other is used simply because the deterministic

Schrödinger equation leads to its use. Nevertheless, consciousness surely plays a very important role for the splittings because of its properties (b) and (c) mentioned above and its *ignorance* of the existence of an infinity of pasts.

6.2 NONLOCALITY

Let us now discuss the *EPR*-like experiments in the *MWI*[5]. Consider then, a pair of photons that are in the correlated state, which in an obvious notation reads (see Eq. (4.15))

$$|\Psi^S\rangle = \frac{1}{\sqrt{2}}(|+-\rangle + |-+\rangle) = |\phi_{+-}\rangle + |\phi_{-+}\rangle \in \mathcal{H}^p \otimes \mathcal{H}^p \equiv \mathcal{H}^S \quad (6.2)$$

So, if one particle has helicity $+1$, the other has helicity -1. Now, suppose that there are two observers $\mathcal{A}$ and $\mathcal{B}$, whose state vectors are

$$|\Psi^\mathcal{A}[\substack{\bullet\bullet\bullet\\\bullet\bullet\bullet}]\rangle \in \mathcal{H}^\mathcal{A}, \quad |\Psi^\mathcal{B}[\substack{\bullet\bullet\bullet\\\bullet\bullet\bullet}]\rangle \in \mathcal{H}^\mathcal{B}, \quad (6.3)$$

before any measurement of the spin projections of the photons is made. In Eq. (6.2), *up* points describe information on the past state of the photon pair and the *down* points describe the state of the other observer. Before the measurement, the state vector of the quantum system plus observers ($\Sigma = \mathcal{S} + \mathcal{A} + \mathcal{B}$) is

$$|\Psi^{S+\mathcal{A}+\mathcal{B}}\rangle = |\Psi^S\rangle \otimes |\Psi^\mathcal{A}[\substack{\bullet\bullet\bullet\\\bullet\bullet\bullet}]\rangle \otimes |\Psi^\mathcal{B}[\substack{\bullet\bullet\bullet\\\bullet\bullet\bullet}]\rangle \in \mathcal{H}^S \otimes \mathcal{H}^\mathcal{A} \otimes \mathcal{H}^\mathcal{B} \quad (6.4)$$

Suppose that a measurement is taken first by observer $\mathcal{A}$, and that the new state vector becomes

$$\begin{aligned}|\Psi^{S+\mathcal{A}'+\mathcal{B}}\rangle &= (|\phi_{+-}\rangle \otimes |\Psi^\mathcal{A}[\substack{\bullet\bullet\bullet\\\bullet\bullet+-}]\rangle \otimes |\Psi^\mathcal{B}[\substack{\bullet\bullet\bullet\\\bullet\bullet\bullet}]\rangle \\ &+ |\phi_{-+}\rangle \otimes |\Psi^\mathcal{A}[\substack{\bullet\bullet\bullet\\\bullet\bullet-+}]\rangle \otimes |\Psi^\mathcal{B}[\substack{\bullet\bullet\bullet\\\bullet\bullet\bullet}]\rangle) \in \mathcal{H}^S \otimes \mathcal{H}^\mathcal{A} \otimes \mathcal{H}^\mathcal{B}.\end{aligned} \quad (6.5)$$

From Eq. (6.5) it is seen that, despite the measurement taken by observer $\mathcal{A}$, observer $\mathcal{B}$ can detects the helicities of the photons as $(+-)$ or $(-+)$ with equal probabilities. There is no wave packet reduction but there is the splitting of $\mathcal{A}$ into two copies; the first sees the photon helicities as $(+-)$ and the other copy sees the photons helicities as $(-+)$. If observer $\mathcal{B}$ measures the helicities of the photons, the new state vector becomes

$$\begin{aligned}|\Psi^{S+\mathcal{A}'+\mathcal{B}'}\rangle &= (|\phi_{+-}\rangle \otimes |\Psi^\mathcal{A}[\substack{\bullet\bullet\bullet\\\bullet\bullet+-}]\rangle \otimes |\Psi^\mathcal{B}[\substack{\bullet\bullet\bullet\\\bullet\bullet-+}]\rangle \\ &+ |\phi_{-+}\rangle \otimes |\Psi^\mathcal{A}[\substack{\bullet\bullet\bullet\\\bullet\bullet-+}]\rangle \otimes |\Psi^\mathcal{B}[\substack{\bullet\bullet\bullet\\\bullet\bullet+-}]\rangle) \in \mathcal{H}^S \otimes \mathcal{H}^\mathcal{A} \otimes \mathcal{H}^\mathcal{B}\end{aligned} \quad (6.6)$$

Now, $\mathcal{B}$ is split into two new copies and as a consequence of all these measurements, we end up with four copies of observer $\mathcal{B}$ seeing all the possibilities. Now, an important role is played by the *meeting* of two observers $\mathcal{A}$ and $\mathcal{B}$. After measuring *one another* (comparison of the state of $\mathcal{S}$), a new wave function arises

$$\begin{aligned}|\Psi^{S+\mathcal{A}''+\mathcal{B}''}\rangle &= (|\phi_{+-}\rangle \otimes |\Psi^\mathcal{A}[\substack{\bullet\bullet\bullet+-\\\bullet\bullet\bullet+-}]\rangle \otimes |\Psi^\mathcal{B}[\substack{\bullet\bullet\bullet-+\\\bullet\bullet\bullet+-}]\rangle \\ &+ |\phi_{-+}\rangle \otimes |\Psi^\mathcal{A}[\substack{\bullet\bullet\bullet-+\\\bullet\bullet\bullet-+}]\rangle \otimes |\Psi^\mathcal{B}[\substack{\bullet\bullet\bullet+-\\\bullet\bullet\bullet-+}]\rangle) \in \mathcal{H}^S \otimes \mathcal{H}^\mathcal{A} \otimes \mathcal{H}^\mathcal{B}\end{aligned} \quad (6.7)$$

62 NONLOCALITY IN QUANTUM PHYSICS

so that it occurs that if observer $\mathcal{A}$ measured photon helicities$(+-)$, for him it is impossible to meet with a copy of $\mathcal{B}$ who measured something different from $(-+)$. Then, we can say that it is because of the identification of $\mathcal{A}$ with only one of its copies that the change of $|\Psi^S\rangle$ into $|\phi_{+-}\rangle$ or $|\phi_{-+}\rangle$ occurs!

If we believe that *everything* is described by the Schrödinger equation, then the identification for one of the copies of $\mathcal{A}$ of the state vector of Σ with

$$|\widetilde{\Psi}^{S+\mathcal{A}''+\mathcal{B}''}\rangle = (|\phi_{+-}\rangle \otimes |\Psi^{\mathcal{A}}[\substack{\bullet\bullet\bullet+-\\ \bullet\bullet\bullet-+}]\rangle \otimes |\Psi^{\mathcal{B}}[\substack{\bullet\bullet\bullet-+\\ \bullet\bullet\bullet+-}]\rangle \quad (6.8)$$

means that it evolved from some *other past state*. But, if the observer *believes* in the existence of a *unique past* he will say that the state vector $|\Psi^{S+\mathcal{A}+\mathcal{B}}\rangle$ fixed in his memory, collapsed into a new state!

In order to have splitting (or multiplication) of observers, some special interaction described by *Hamiltonian measurement* is necessary. For von Neumann measurements (when the state vector after measurement is in an eigenstate of the measurable observable), this means that if before the interaction there is some state $|\Psi^{S+\mathcal{A}}\rangle = |\Psi^S\rangle \otimes |\Psi_0^{\mathcal{A}}\rangle$, where $|\Psi_0^{\mathcal{A}}\rangle$ is the initial state vector of the observer, then after the measurement, if $|\Psi_0^{\mathcal{A}}\rangle = c_1|1\rangle + c_2|2\rangle$, $|\Psi^{S+\mathcal{A}}\rangle$ evolves to

$$c_1|\Psi_1^{\mathcal{A}}\rangle \otimes |1\rangle + c_2|\Psi_2^{\mathcal{A}}\rangle \otimes |2\rangle. \quad (6.9)$$

In this *realistic* interpretation the properties of quantum systems exist before measurements, because if the wave function is fixed, then it is an eigenfunction of the operator corresponding to some observable. Nevertheless, we have a breakdown of Bell's inequalities, because due to the infinite number of pasts, the *relevant* copy of the observer has in his memory a past, which is different from that, which evolves in observable results!

Some authors think that the existence of many universes may be naturally connected with the quantization of gravity, leading to the existence of many different geometries. The *MWI* could lead to some differences with the standard quantum theory. It is possible to conjecture that the different universes do not interact *now* because gravity is almost classic, but in the quantized earlier universe(s) there is an interaction between the different branches corresponding to different geometries. In this respect, see Hawking who studied the possible role of wormholes connecting different universes due to a tunneling effect[6].

6.3 CRITICISMS

The main *criticisms* of the *MWI* are the following:

(i) The supposition that *actual infinity* of different universes where every possible situation occurs in one of them, does not have an adequate mathematical formalism. We can imagine some system (the many universes) with an infinite number of degrees of freedom evolving in time due to the Schrödinger equation. But, what is the meaning (if any) of the parameters, discriminating each one of the universe's copies, what is the general Hamiltonian of evolution? All this is absolutely unclear at present.

(ii) The problem of the *preferable basis*. Different from the observer dependent Copenhagen interpretation, where it is the observer who makes the choice of some *preferable* basis when he chooses to measure some observable in order to get information about the quantum object, in the *MWI* there is a real problem. Indeed, suppose that the initial situation of the quantum system plus observer is described by $|\Psi^S\rangle \otimes |\Psi_0^A\rangle$ and during the measurement it evolves into some new state $|\chi^{S,A}\rangle$. We can write many different representations of this state in terms of different orthonormal bases, different from, let us say, $c_1|\Psi_1^A\rangle \otimes |1\rangle + c_2|\Psi_2^A\rangle \otimes |2\rangle$. Why would nature prefer one basis instead of others?

(iii) The idea of the observer's splitting can lead to different values for probabilities of the observables, different from the ones given by Born's postulate. Indeed, consider the Stern-Gerlach experiment in which a beam of particles is splitted into two other beams with spin projections $S_z = +1/2$ and $S_z = -1/2$. Suppose that the initial sate vector of the beam is such that the probability for the obtention of $S_z = +1/2$ is $1/3$. But the observer is split into two observers, and each copy has equal status. Then, it is natural to have probability 1/2 in a *realistic picture* (where classical probability is valid). But, the frequencies are given by the quantum formula, which contradicts the result obtained through the above reasoning. In [7] the interested reader can also find an attempt to solve the problem of existence of many microworlds with one macroworld.

We mention also that there are attempts by Squires[8] and d'Espagnat[9] to understand the *MWI* as a *many points of view interpretation*. Their reasoning is:

Why must we think about different worlds as *real* and not *potential*? Taking the postulate of only one real world with one past, we can call the other universes potentialities or points of view of the copies of the observer, which are to be intended not as real, but as *imaginable* copies. In such a view we come to something that looks like the von Neumann, London and Bauer interpretation, discussed in the previous Chapter. The only difference has to do with the claim that somehow the Schrödinger's evolution describes the world of *potentialities*, but that the observer, due to his introspection, creates the eventual reality by his decision of identification of himself with one of the many potentialities.

At last, we would like to comment that the existence of many pasts is similar to the existence of many futures defined by the present if free will exists. However, we must stress here that a special role is played by the *present* and the *memory* in this scheme. *Present* data can be explained by evolution from *many pasts*.

Concerning memory we can ask: why doesn't the observer, jumping from one world to another, forget previous experiences, when he is aware of an evolving wave function which was prepared into a superposition?

The reason is that if he loses his memory, then no splitting can be observed, but the Schrödinger equation which often leads to the evolution of an eigenstate of some operator into a superposition of its eigenvalues will be violated for this observer. There must be a crucial difference between a today's present (i.e., the nowhere, for each observer) and the past and and future. There are many worlds in both past and future, but only one world at present where all these worlds meet...Memory is some data at present.

So, *now* each one of us can have some darkened photoplate from our past experience together, let us say, with some new observation of the state of a given quantum particle, showing the leap of the wave function[1].

Bell's many pasts idea, leading to one present seems to be the best interpretation of the MWI for the following reason: We recall that in Laplace determinism, there is always one equation of motion with unique initial conditions on the Cauchy surface, and there is no difference between present, past and future. With Bell's idea, the present data evolved due to the Schrödinger equation from data that in the past are contradictory, and can have a meaning only if there are many pasts. Nevertheless, there is still a two worlds picture: for classical phenomena like, e.g., the motion of the Moon, we have only one past, but for quantum systems there are many pasts. This is due to the existence of complementarity — noncommuting operators corresponding to the observables of quantum systems and to the absence of noncommuting observables for classical systems (again our old Schrödinger's cat paradox).

We end this Chapter with the observation that a useful collection of articles of the MWI can be found in [10].

Notes

1. One of the authors, A. A. Grib, discussed in 1968[7] the possibility of a many vacua world as microworlds within one macroworld, where similar problems arose.

References

[1] H. Everett, "Relative state" formulation of quantum mechanics, *Rev. Mod. Phys.* **29**(3), 454–462 (1957).

[2] J. A. Wheeler, Assessment of Everett's 'relative state' formulation of quantum theory, *Rev. Mod. Phys.* **29**(3), 463–465 (1957).

[3] B. S. DeWitt, in: *The Battelle Rencontres, I, 1967, Lectures in Mathematical Physics*, edited by C. DeWitt and J. A. Wheeler (W. A. Benjamin, New York, 1968), pp. 318–332.

[4] J. S. Bell, *Quantum Mechanics for Cosmologists*, in: *Quantum Gravity 2*, edited by C. Isham, R. Penrose and D. Schiama (Clarendon Press, Oxford 1982), pp. 611–637.

[5] M. Bitbol, An analysis of of the Einstein-Podolski-Rosen correlations in terms of events, *Phys. Lett. A* **96**(2), 66–70 (1983).

[6] S. Hawking, Baby universes, *Modern Phys. Lett A* **5**(2), 145–155 (1990).

[7] A. A. Grib, Breaking of CP invariance in K-meson decay and nonequivalent representations in quantum field theory (in Russian), *Vestnik of the Leningrad University* **22**, 50–56 (1967), (Izdatelstvo Leningradskogo Universiteta).

[8] E. J. Squires, History and many worlds quantum theory, *Found. Phys. Lett.* **5**(3), 279–290 (1992)

[9] B. d'Espagnat, *Conceptual Foundations of Quantum Theory* (W. A. Benjamin, Reading, MA, 1976).

[10] B. S. DeWitt and N. Graham, *The Many Worlds Interpretation of Quantum Mechanics* (Princeton University Press, Princeton, New Jersey, 1973).

7
QUANTUM LOGIC (QL)

7.1 PROPOSAL

There are many tentatives to apply non classical logics to quantum theory (see, e.g., the excellent book of Jammer[1]). Here we will be concerned with non distributive quantum logic, (which we will abbreviate as *QL*, in what follows) first introduced by Birkhoff and von Neumann[2] and which has been developed by Jauch[3], Piron[4], Finkelstein[5,6], Grib[7-9], Zapatrin [10-11].

It is opportune to present, even if very briefly, how the idea of *QL* appeared. First we recall that, as is well known, to Aristotle's formal logic (with some improvements, see, e.g., Putnam[12]), we can give an axiomatic presentation that is known as proposition calculus. It is very important to consider that the calculus of formal logic has been obtained by analyzing the meaning of the intrinsic relations between propositions according to categories of perception and judgment of the human mind(s) and so must be considered (as emphasized by Jauch[13]) as tautologically true in all circumstances. In 1847, Boole discovered an algebraic structure[1], now known by the name of Boolean algebra, which is isomorphic to the proposition calculus. Since human beings reason[2] using formal logic, we will use sometimes the designation of *Boolean observers* for them.

Now, every measurement of a physical system can always be reduced to a specific kind of measurement called *yes-no* experiments (*YNE*), i.e., observations that permits as an answer only one of two possible alternatives. The basic idea is that the results of a sufficient number of *yes-no* experiments permit us to determine the physical properties of the physical system and the *state*[3] of the system after its preparation. Jauch[13] calls a *YNE* that can select the values of a measurable physical quantity, a *filter*. If γ is a given *YNE*, we denote by γ^c the *YNE* done with the same equipment and such that the result of the measurement of γ is yes (respectively no), then the result of the measurement of γ^c is no (respectively yes).

The set of all *YNE* is a poset (see Appendix A). To see this, first define a *YNE* as being *true* for a given state of a physical system if its measurement will give the result yes with probability one, i.e., with certainty. Now, it is a fact of experience that certain pairs of *YNE*, let us call them γ and δ, are such that when γ is true then δ is also true. Writing this fact by $\gamma \leq \delta$ it becomes a trivial exercise to verify that the set of all *YNE* is indeed a poset.

A *YNE* β is said to be *equivalent* to a *YNE* α and we write $\alpha \sim \beta$, if and only if

$$\alpha \leq \beta \text{ and } \beta \leq \alpha. \tag{7.1}$$

Let α be a *YNE* and let $\{\alpha_i\}$ be an indexed family of *YNE* such that $\alpha \sim \alpha_i, \forall i \in I$ (the index set). The equivalence class of *YNE*, i.e., the set

$$a = \{\alpha_i | \alpha \sim \alpha_i\} \tag{7.2}$$

will be called a *proposition* of the physical system.

Let $a = \{\alpha_i | \alpha \sim \alpha_i\}$ and $b = \{\beta_j | \beta \sim \beta_j\}$ be arbitrary propositions of a physical system. We will write

$$a \leq b \tag{7.3}$$

if $\alpha_i \leq \beta_j, \forall \alpha_i \in a, \forall \beta_j \in b$. The physical meaning of Eq. (7.3) is obvious.

Let $\{\gamma_i\}$ be an indexed family of *YNE*. From the set $\{\gamma_i\}$ we define a new YNE denoted $\Pi \gamma_i$ which is to be physically obtained by a random choice of *any* one of the equipments used in the determination of any γ_i and then record the outcome as the result of the measurement of $\Pi \gamma_i$. With this definition we always have

$$(\Pi \gamma_i)^c = \Pi \gamma_i^c. \tag{7.4}$$

The equivalence class of the *YNE* $\Pi \gamma_i$ will be denoted in what follows by $\wedge_{i \in I} \gamma_i$ or simply $\wedge \gamma_i$.

It is easy to verify that if all *YNE* γ_i are true, then $\Pi \gamma_i$ is true. More important, $\wedge_{i \in I} \gamma_i$ has the property that in Appendix A we have called the *greatest lower bound* (or *meet*). Indeed, let $\{a_j\}, j \in I$, be the set of all propositions such that

$$a_j = \{\alpha_{ji} | \alpha_{ji} \sim \gamma_i\}. \tag{7.5}$$

Then, given a proposition b we have

$$b \leq (\wedge_{j \in I} a_j) \Leftrightarrow b \leq a_j, \forall j \in I. \tag{7.6}$$

Write

$$\vee_{j \in I} a_j = \wedge_{a_j \leq x} x. \tag{7.7}$$

We can then verify that $\vee_{j\in I}a_j$ has all the properties of a *least upper bound* (or *joint*) for the set of propositions $\{a_i\}, j \in I$, as defined in Appendix A, since given a proposition b we have

$$\vee_{j\in I}a_j \leq b \Leftrightarrow a_j \leq b, \forall j \in I. \tag{7.8}$$

If $\mathcal{L}$ is the set of all propositions of a given physical system, then it is clear that the proposition $\wedge_{a\in\mathcal{L}}a$ must be always false. We denoted by $\mathbf{0}$ the *false* (or trivial, or absurd) proposition and write

$$\mathbf{0} = \wedge_{a\in\mathcal{L}}a. \tag{7.9}$$

Also, the proposition $\vee_{a\in\mathcal{L}}a$ is always true. It is the trivial proposition which will be denoted by $\mathbf{1}$. We write

$$\mathbf{1} = \vee_{a\in\mathcal{L}}a. \tag{7.10}$$

Given a proposition a by definition a *complement* of a is a proposition b such that

$$a \wedge b = \mathbf{0} \text{ and } a \vee b = \mathbf{1}. \tag{7.11}$$

It is obvious that a given proposition can have many complements. We say that b is a *compatible complement* of a and denote it by a', if there exists a *YNP* $\alpha \in a$ such that $\alpha^c \in b = \{\beta_i | \beta_i \sim \beta\}$.

We define the *state* of a physical system (be it classical or quantum) whose set of all propositions is $\mathcal{L}$ as being the set $\mathfrak{S} \subseteq \mathcal{L}$ such that

$$\mathfrak{S} = \{a \in \mathcal{L} | a \text{ is } true\}. \tag{7.12}$$

All *known* physical systems are such that every possible proposition $a \in \mathcal{L}$ has at least one compatible complement $a' \in \mathcal{L}$. In the following, this empirical evidence is taken as an *axiom* satisfied by all physical systems. It will be called the *compatibility axiom*.

As we just saw, the set of all propositions $\mathcal{L}$ of any given physical system (be it classical or quantum) possesses some structural properties that are *independent* of the state of the system. Recalling Appendix A (definition 31), we realize that the set of all propositions with the above properties define a *complete lattice*[15], this being the reason for the notation $\mathcal{L}$.

The lattice $\mathcal{L}$ is also atomic (definition 32 of Appendix A) since there are atoms in it, i.e., elements $a \neq \mathbf{0}$ for which there is no $b \neq \mathbf{0}$, such that $\mathbf{0} \leq b \leq a$. The atoms are the elementary propositions and every non-zero element of the lattice is either an atom, or there is an atom b, such that $\mathbf{0} \leq b \leq a$.

In fact, the most general lattice of propositions of a physical system is a *complete orthocomplemented and weakly modular lattice* (see definitions 47 and 49 of Appendix A). This can be seen with the introduction of the Piron axiom[3,16] and the concept of compatible and incompatible propositions.

For all known physical systems if $a, b \in \mathcal{L}$, and if $a \leq b$, then the sublattice $\{a, b, a', b'\} \subseteq \mathcal{L}$ is a Boolean lattice. In the following, this empirical evidence is taken as an axiom satisfied by all physical systems and will be called the *Piron axiom*[3]. With the help of the compatibility and Piron axioms we can easily prove the following:

proposition. The mapping

$$' : \mathfrak{L} \ni a \mapsto a' \in \mathfrak{L}, \tag{7.13}$$

is *unique* and corresponds to what we called in Appendix A (definition 47) *orthocomplementation*. Thus, we have for $\forall a, b \in \mathfrak{L}$ that

$$a \leq b \Rightarrow b' \leq a', (a')' = a. \tag{7.14}$$

7.1.1 WEAK MODULARITY POSTULATE AND PROPOSITION CALCULUS

We now introduce the mathematical definition of *compatible* and *incompatible* propositions. The definition to be given below is to mirror the empirical fact that compatible propositions can be measured simultaneously. In classical physics all propositions are compatible and as we are going to see in detail in the next section they define a Boolean lattice. We proceed as follows:

Let $a, b \in \mathfrak{L}$ be two propositions. We say the propositions a, b, are *compatible* if the set $\mathfrak{S} = \{a, b\}$ generates (see definition 40 of Appendix A) a Boolean sublattice. When a and b are compatible we write that $a \leftrightarrow b$. The notation emphasizes the fact that compatibility is a *symmetrical* relationship. However it is important to realize that compatibility is not *transitive*, i.e., if $a \leftrightarrow b$, $b \leftrightarrow c$, in general it is not true that $a \leftrightarrow c$. Now since when a and b are compatible the set $\{a, b\} = \mathfrak{S}$ generates a Boolean sublattice, then the distributive laws (Eq. (A.15)) valid in such lattices imply that

$$(a \wedge b') \vee b = (b \wedge a') \vee a \tag{7.15}$$

To develop a proposition calculus able to encode besides the laws of classical theory, also the laws of quantum theory, we need to introduce the *WMP*.

Weak Modularity Postulate[4] (WMP): If $a \leq b$ then $a \leftrightarrow b$.

The *center* of $\mathfrak{L}$ (definition 68 of Appendix A) is the set $\mathfrak{C}$ of all propositions that are compatible with all propositions of $\mathfrak{L}$. It is obvious that $\mathfrak{C}$ is a Boolean lattice and that **0** and **1** are compatible with all propositions of the lattice. When $\mathfrak{C} = \{\mathbf{0}, \mathbf{1}\}$, then it is said to be trivial and the lattice is said to be irreducible. If $\mathfrak{C} \neq \{\mathbf{0}, \mathbf{1}\}$, then the lattice is said to be reducible. We now define that a subset $\mathfrak{G} \subset \mathfrak{L}$ is said to be a *compatible* set of propositions if and only if the lattice generated by $\mathfrak{S}$ is a Boolean sublattice of $\mathfrak{L}$. If a given set of propositions does not define a Boolean sublattice, the set of properties is said to be *incompatible*.

The lattice of propositions $\mathfrak{L}$ (for a given physical system, classical or quantum) equipped with the weak modularity postulate is a complete orthocomplemented and weakly modular lattice which is said to be the *proposition calculus* of the physical system.

It is fundamental to have in mind that the structure of the lattice $\mathfrak{L}$ is independent of the state of the physical system, be it a classical or a quantum system. We will say that the lattice describes the *intrinsic structure* of the physical system. In this sense it describes the set of *qualities* defining the physical system.

7.1.2 THE BOOLEAN LATTICE OF A CLASSICAL SYSTEM

At this point, it is essential to distinguish between *classical* and *quantum* systems. For classical systems, it is always possible to measure the proposition $a \wedge b$ if a and b can be measured separately. This possibility is due to the assumption that for a classical system any measurement can be taken without disturbing the system. Then, measuring first a or b does not change the physical state of the system. For quantum systems we know that this assumption is not true and the meaning of the proposition $a \wedge b$ is given by the definition of the greatest lower bound (definition 27, Appendix A). This implies that if we measure, e.g., a first, we are not sure if the measurement hasn't changed the state of the system in such a way as to affect the measurement of b. Also, after measuring b, we are not able to say that the system still satisfies proposition a. Following Jauch, we said that $a \wedge b$ is true if the system is a filter for the proposition $a \wedge b$. This filter is "made" of an infinite sequence of alternating pairs of filters.

As is well known, the possible states of a classical system are described by the points $s \equiv (\vec{q}, \vec{p}) \in \mathcal{M}_{qp}$, the phase space of the system and any physical measurable quantity of the system can be represented by a real Borel function (definition 80 of Appendix A),

$$F : \mathcal{M}_{qp} \to \mathcal{R}. \tag{7.16}$$

Then, the elementary propositions associated with F are the propositions

$$F(\vec{q}, \vec{p}) \in \mathcal{J} \subset \mathcal{R} \tag{7.17}$$

where $\mathcal{J}$ is a Borel set (definition 77 of Appendix A). To each proposition of this kind it is possible to associate a unique Borel set in $\mathcal{M}_{qp}$, by

$$F^{-1}(\mathcal{J}) \equiv \{\vec{q}, \vec{p} : F(\vec{q}, \vec{p}) \in \mathcal{J} \subset \mathcal{R}\}. \tag{7.18}$$

Also, to any Borel set $\mathcal{S} \subset \mathcal{M}_{qp}$, we can associate a proposition called *elementary proposition*. Recall that the *characteristic function* of $\mathcal{S}$ is

$$\chi_{\mathcal{S}} : \mathcal{M}_{qp} \supset \mathcal{S} \to \mathcal{R}, \; \chi_{\mathcal{S}}(\vec{q}, \vec{p}) = 1. \tag{7.19}$$

In this way, every Borel set $\mathcal{S} \subset \mathcal{M}_{qp}$ represents indeed a *proposition* of the physical system and is true if $(\vec{q}, \vec{p}) \in \mathcal{S} \subset \mathcal{M}_{qp}$. It is obvious that the absurd proposition $\mathbf{0}$ is represented by the empty set and that the trivial proposition $\mathbf{1}$ is represented by $\mathcal{M}_{qp}$.

Defining for a classical system the set of all propositions (i.e., the class of Borel sets of $\mathcal{M}_{qp}$, denoted by $\mathfrak{L}_c$), the ordering relation by the set theoretical inclusion, the meet by set theoretical intersection and the joint by the set theoretical union and recalling the definition 40 of Appendix A, we realize that this set is a Boolean lattice or more precisely a σ-Boolean lattice, for which the so called distributive laws hold, i.e., if $\mathfrak{a}, \mathfrak{b}, \mathfrak{c} \in \mathfrak{L}_c$, then

$$\mathfrak{a} \cap (\mathfrak{b} \cup \mathfrak{c}) = (\mathfrak{a} \cap \mathfrak{b}) \cup (\mathfrak{a} \cap \mathfrak{c}),$$
$$\mathfrak{a} \cup (\mathfrak{b} \cap \mathfrak{c}) = (\mathfrak{a} \cup \mathfrak{b}) \cap (\mathfrak{a} \cup \mathfrak{c}). \tag{7.20}$$

It is really an interesting fact that the set of propositions of a classical system has the same mathematical structure as the proposition calculus of classical formal logic, where the meet $\wedge$ corresponds to conjunction — *and* — and the joint corresponds to the disjunction — *inclusive or*. This is even more significant if we recall that: (a) according to what we said above, the set of propositions may be identified with the set of qualities that define a given classical physical system in the class of all classical physical systems; (b) that there exists in formal logic an isomorphism between the calculus of the intrinsic logic relationships of qualities of objects in a given class and the calculus of intrinsic logic relationships of propositions.

It is *"as if there exists an adequacy between the intellect and things"* to use a well known statement of Isaac Israeli, a tenth century philosopher. The *adequacy* is interpreted by saying that classical systems have qualities that follow Boolean logic.

7.1.3 THE NON DISTRIBUTIVE LATTICE ASSOCIATED TO A QUANTUM SYSTEM

We now show that the set of propositions of a physical system that follows the laws of quantum theory as presented in Chapter 2 is a non distributive complete and orthocomplemented weakly modular lattice, which is modular, (i.e., $a, b, c \in \mathfrak{L}$ then $a \geq c \Rightarrow a \wedge (b \vee c) = (a \wedge b) \vee c$)) only for quantum systems whose possible states belong to a finite dimensional Hilbert space.

Let $\mathcal{H}$ be the Hilbert space of a quantum system and let $\{\mathcal{H}_\alpha\}$, $\alpha \in J$ an indexed family of subspaces of $\mathcal{H}$.

By *definition* every proposition of the quantum system described by $\mathcal{H}$ is some set of subspaces of the indexed family $\{\mathcal{H}_\alpha\}$. Equivalently (as it is clear from our discussion in Appendix B) every proposition can be also defined by the corresponding set of the indexed family $\{\mathbf{P}_\alpha\}$ of projection operators.

With the above definition and recalling proposition 32 of Appendix A that says that the set $\mathfrak{L}(\mathcal{H})$ of the subspaces of $\mathcal{H}$ (or $\mathfrak{L}^\mathbf{P}(\mathcal{H})$), the set of the corresponding projection operators)[5] is a complete orthocomplemented and non modular lattice with appropriate identifications of the operations $\leq, \wedge, \vee$ and $'$ with operations in the set of subspaces.

In details we have:

(a) we interpret the *embedding* of $\mathcal{H}_a$ in $\mathcal{H}_b$ ($\mathcal{H}_a \subseteq \mathcal{H}_b$) as the order relation $\mathcal{H}_a \leq \mathcal{H}_b$.

We know that in this case the corresponding projectors $\mathbf{P}_a$ and $\mathbf{P}_b$ satisfy the condition $\mathbf{P}_a \mathbf{P}_b = \mathbf{P}_a$ with the obvious physical interpretation that says that if the proposition $\mathbf{P}_a$ is true, then $\mathbf{P}_b$ is also true.

(b) we interpret the set theoretical intersection ($\cap$) of subspaces $\mathcal{H}_a, \mathcal{H}_b \subseteq \mathcal{H}$ (which is the largest subspace contained both in $\mathcal{H}_a, \mathcal{H}_b \subseteq \mathcal{H}$) as the operation $\wedge$ in the abstract definition of a lattice.

(c) the operation ∨ in an abstract lattice is represented by the *algebraic union* of subspaces, denoted by ⊎ (definition 15 of Appendix B). We write the algebraic union of two subspaces $\mathcal{H}_a, \mathcal{H}_b \subseteq \mathcal{H}$ as $\mathcal{H}_a \uplus \mathcal{H}_b$.

(d) we interpret "'" in an abstract lattice as being the operation of taking the orthogonal subspace of a given space, and identify $|0\rangle$ (the zero vector of $\mathcal{H}$) and $\mathcal{H}$, respectively as the elements **0** and **1** of the lattice. We can then verify that if $\mathcal{H}_a$ is a subspace of $\mathcal{H}$, then $\mathcal{H}_a^\perp$, the orthogonal subspace of $\mathcal{H}_a$ is also in $\mathfrak{L}(\mathcal{H})$, and $\mathcal{H}_a \cap \mathcal{H}_a^\perp = |0\rangle$, and $|0\rangle^\perp = \mathcal{H}$. It follows that the lattice always contains $|0\rangle$ and $\mathcal{H}$, as it may be the case.

We recall from example 29 of Appendix B, which the reader is invited to study before continuing, that the lattice of propositions of a quantum system is *non distributive*.

If we denote by $\mathcal{H}_a^c$ the *set theoretical complement* of $\mathcal{H}_a \subseteq \mathcal{H}$, then we have immediately that given any pair of sets, $\mathcal{H}_a, \mathcal{H}_b \subset \mathcal{H}$,

$$(\mathcal{H}_b \cap \mathcal{H}_a) \cup (\mathcal{H}_b \cap \mathcal{H}_a^c) = \mathcal{H}_b, \qquad (7.21)$$

which follows immediately from the distributivity laws associated with the operations ∩ and ∪ in the Boolean algebra of sets. Now, we have already seen above that in $\mathfrak{L}(\mathcal{H})$ it is the operation of taking the orthogonal complement of a given subspace the one that corresponds to the operation "'" in the definition of an abstract lattice.

This suggests the definition: two subspaces $\mathcal{H}_a, \mathcal{H}_b \subseteq \mathcal{H}$ are *compatible* if

$$(\mathcal{H}_b \cap \mathcal{H}_a) \uplus (\mathcal{H}_b \cap \mathcal{H}_a^\perp) = \mathcal{H}_b. \qquad (7.22)$$

We recall that there are other more convenient ways for expressing the compatibility relation. See proposition 27 of Appendix B.

As can be easily verified using proposition 27 of Appendix B, this relation is symmetric under the interchange between $\mathcal{H}_a$ and $\mathcal{H}_b$, which suggests writing the compatibility relation as $\mathcal{H}_a \leftrightarrow \mathcal{H}_b$. With this definition $\mathfrak{L}(\mathcal{H})$ (or $\mathfrak{L}^\mathbf{P}(\mathcal{H})$) becomes a weakly modular lattice (definition 49 of Appendix A)

We recall here proposition 28 of Appendix B that says that three subspaces $\mathcal{H}_a, \mathcal{H}_b, \mathcal{H}_c \subseteq \mathcal{H}$ are compatible if they satisfy the distributive laws,

$$\begin{aligned}\mathcal{H}_c \cap (\mathcal{H}_a \uplus \mathcal{H}_b) &= (\mathcal{H}_c \cap \mathcal{H}_a) \uplus (\mathcal{H}_c \cap \mathcal{H}_b), \\ \mathcal{H}_c \uplus (\mathcal{H}_a \cap \mathcal{H}_b) &= (\mathcal{H}_c \uplus \mathcal{H}_a) \cap (\mathcal{H}_c \uplus \mathcal{H}_b).\end{aligned} \qquad (7.23)$$

This ends our identifications for the basic operations of the set of propositions of a quantum system when this set is represented by $\mathfrak{L}(\mathcal{H})$. For the identifications for the basic operations of the set of the propositions when it is represented by $\mathfrak{L}^\mathbf{P}(\mathcal{H})$, the reader is invited to consult section B.5 of Appendix B.

What we just *learned* is that contrary to what happens for the case of classical systems, for quantum systems, the associated proposition calculus is a non Boolean lattice. It is quite clear that the lattice $\mathfrak{L}(\mathcal{H})$ is not a logic in the philosophical sense, but nevertheless, we remark that if we say that the elements of the set $\mathfrak{L}(\mathcal{H})$ define the qualities of a system, then in this case there is "*no adequacy between the intellect and things.*"

This last remark, suggests that a *radical* interpretation of *QL* in the spirit of Finkelstein[5,6] and Putnam[17][6] can be given. First, we remark that *QL* can be thought of as some version of the *CI*, because by using the mathematical structure of a non Boolean lattice we can obtain[18] all the mathematical theory of the Hilbert spaces and their operators as presented in Chapter 1. Second, the observer also plays an important role, but here the *claim* is that there exists an external quantum reality outside the observer, which possesses a logic different from the Boolean logic of the observer.

Wave packet collapse occurs due to the necessity that the observer has to make his intellect adequate to "things", i.e., to project non-human, non Boolean logic of quantum systems onto his standard Aristotelian logic in order to get information about the quantum reality.

From this point of view, the discovery of quantum theory is the discovery of a new logic. Microscopic systems follow *QL* , and *almost* all macroscopic classical systems follow Boolean logic. We say almost because if the external reality is a quantum reality, then we can ask: are there realizations of *QL* for macroscopic systems? The answer is *yes* and in the next Chapter we shall present some examples.

We close this section by discussing some standard quantum mechanical examples using *QL*.

(e.i) Suppose that we have an electron for which the spin projections on two axes, x and y are measured. Then if we obtain $S_x = 1/2$ and $S_y = -1/2$ as the manifestations of the spin projection properties at two different times, this does not mean that if we measure S_x again, we will necessarily obtain the value 1/2. The absence of simultaneously definite values of S_x and S_y, according to *QL* means that:

$$S_x = 1/2 \wedge (S_y = 1/2 \vee S_y = -1/2) \text{ is true} \qquad (7.24)$$

but,

$$(S_x = 1/2 \wedge S_y = 1/2) \vee (S_x = 1/2 \wedge S_y = -1/2) \text{ is false.} \qquad (7.25)$$

So, in the *QL* interpretation, we can say that the properties of quantum objects exist independently of the observer, but if we interpret, respectively the meet $\wedge$ and the join $\vee$ as meaning conjunction *and*, disjunction *or*, then the *semantical* meaning of these words is not the usual one. We can say if we are *empiricists* that we just discovered the empirical meaning of *and* and *or*. The empirical meaning is more general than the one we inferred from our immediate experience with classical systems whose propositional calculus is a Boolean lattice. As we will see through the examples below, for the intelligibility of the empirical *and* and *or* of *QL*, the observer must *invent* time.

(e.ii) In the double slit experiment we can say according to *QL* that a particle with definite momentum has a coordinate which is x_1 or x_2 (coordinates of the holes in the screen) *not identical to either* x_1 or x_2 . So, in *QL or* is not as different from *and* as it is in standard Boolean logic. That is why people sometimes say that: *an electron, being a pointlike particle passes simultaneously through both holes in the screen.*

7.2 HASSE DIAGRAMS

To become familiar with the properties of *QL*, it is very useful to draw some so-called *Hasse diagrams* for some simple quantum systems. A Hasse diagram of a lattice $\mathfrak{L}$ is a graphical representation of the lattice where for $\forall a, b \in \mathcal{L}$ if $a \leq b$, we plot a lower than b and then we represent the implication relation (i.e., the partial order relation) between a and b by more or less vertical lines.

7.2.1 SPIN ONE SYSTEM FOR WHICH TWO SPIN PROJECTIONS, S_X AND S_Z, ARE MEASURED

In Fig.(7.1), 1, 2, 3, 4, 5, 6, correspond to properties of the system that are independent from one another. We can get information about this system by using appropriate measuring devices. We choose as our measuring apparatus in the example, one that has a series of lamps disposed in the configuration of the lattice. For example, if the system has the property 1, the electric lamp 1 is on and all the others are off. If 2 is true, then lamp 2 is on and all the others are off, etc...

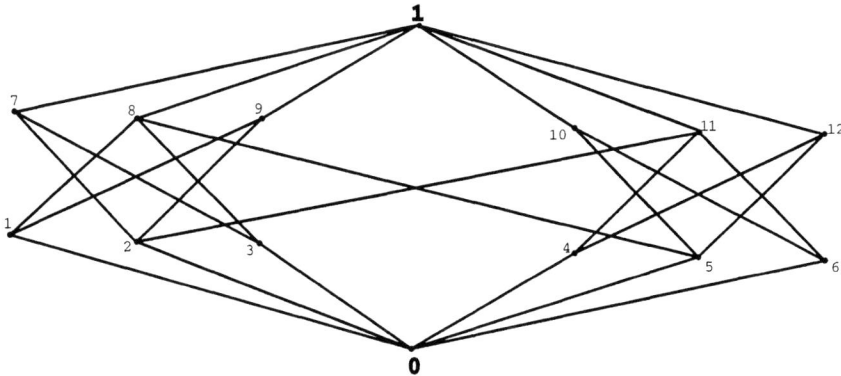

Figure 7.1. Hasse diagram for a spin one system for which S_x and S_z are measured.

Let us then write,

$$1 \wedge 2 = 1 \wedge 3 ... = 1 \wedge 6 = 2 \wedge 3 = ... = 5 \wedge 6 = 0, \qquad (7.26)$$

i.e., the lamps cannot light simultaneously (**0** is always false, **1** is always true).

We can *call* the event, lamp 7 on, as the event 2 or 3 ($2 \vee 3$), because if we understand the lines in our diagram as *connections*, then, lamp 7 is on each time 2 is on or 3 is on. Thus, the event lamp 3 is on, can be understood as $(7 \wedge 8)$, because if 3 is on, then both lamps 7 and 8 connected with it by lines (wires), are on. Lamp 1 is on when all the lamps are on.

It is easy to see that our lattice is not distributive. Indeed,

$$1 \wedge (3 \vee 4) = 1 \wedge \mathbf{1} = 1 \neq (1 \wedge 3) \vee (1 \wedge 4) = \mathbf{0} \wedge \mathbf{0} = \mathbf{0}. \qquad (7.27)$$

It follows that for a non distributive lattice we cannot define the classical probability measure. Instead, we must use the so called *probability amplitudes* (represented by the state vector $|\Psi\rangle$). To each element of our lattice corresponds some projector $\mathbf{P}_L$

76   NONLOCALITY IN QUANTUM PHYSICS

and if $|\Psi\rangle$, the state vector is known, we can define:

$$\mu(L) = \langle\Psi|\mathbf{P}_L|\Psi\rangle \qquad (7.28)$$

as giving the probabilities according to Born's rule for a yes answer concerning the property L.

As it is easy to see, there are distributive triplets, $(1,2,3), (4,5,6)$. For them, it holds

$$\begin{aligned} 1 \wedge (2 \vee 3) &= (1 \wedge 2) \vee (1 \wedge 3) \\ 4 \wedge (5 \vee 6) &= (4 \wedge 5) \vee (4 \wedge 6) \end{aligned} \qquad (7.29)$$

It is only when we take one of the atoms from the left side and the other from the right side, that we get nondistributivity. To the elements of sets for which the distributivity law does not hold, there correspond, as we saw above, noncommuting operators in the Hilbert space of the quantum system. To elements of the sets for which the distributivity law holds, there correspond commuting operators. In our diagram, atoms 1,2,3 correspond to $S_z = +1, 0, -1$; to the atoms 4,5,6 correspond $S_x = +1, 0, -1$.

Due to QL it is possible to say that the object has spin projection S_x equal to 1 (atom 1) *and* spin projection S_z equal to $+1$ or 0 or -1, because

$$1 = 1 \wedge (4 \vee 5 \vee 6) \qquad (7.30)$$

It is false, however, to say that at the same *moment* (as we say when we use the classical understanding of *and*) the object has x and z spin projections equal to 1 and 1 *or* $+1$, 1 *or* 0, 1 *or* -1.

7.3   THE BIRTH OF TIME

The role of the observer in QL interpretation is inferred from the non *adequacy* between his Boolean distributive logic (which is also *materialized* for the classical systems that he uses as measuring devices) and the non Boolean logic of the quantum world. The *result* of this inadequacy is the wave packet collapse.

Indeed, the non Boolean nondistributive lattice is not isomorphic to the Boolean logic of the observer. The conscious observer solved (long ago) the problem of the *adequation* of the Boolean structure of his logic and the non Boolean logic of a large class of phenomena that occur outside his mind. How? By inventing a *new quality* for the relations between these phenomena. This new quality is represented by a parameter—*time*.

Let us see how with the invention of time, any contradiction that an observer might find between his Boolean logic and the non Boolean world, disappears.

Referring again to our very simple quantum system described by the Hasse diagram of Fig.(7.1), we observe the following:

For the Boolean observer, if 1 is true and if it is equal to $1 \wedge (4 \vee 5 \vee 6) = 1$, then $(4 \vee 5 \vee 6)$ is true, and due to his Boolean structure, either 4 or 5 or 6 must be true. In non-Boolean logic it is possible, as we already know, to have 4-false, 5-false, 6-false, but nevertheless (4 ∨ 5 ∨ 6)- true. But, for the Boolean observer this is impossible! So, he will say that at some *moment of time*, one of the 4,5,6 (totally undetermined, and this is the source of quantum indeterminism) *becomes* true[23]. So, *becoming* or *movement in time* appears because of the difference of the two logics!

An observer always measures noncommuting observables at *different* moments of time and it is impossible for him to get information about them simultaneously.

We can consider this profound difference in the logical structures of the quantum world and the logic of our consciousness as the *reason* why, we as human beings always move in time to the future, while in space (which must be interpreted also as a kind of *relation* needed to accommodate the properties of phenomena that are outside the observer's mind), we can be at rest in a given point. This happens only due to the fact that the observer can *identify* himself as some union of a Boolean consciousness and a material body (consisting of quantum objects), only if he is *moving in time*!

To the Boolean sublattices of the non-Boolean lattice, the observer can give the interpretation in terms of events in Minkowski spacetime[7].

With the above remarks, we see then, that the interpretation of *QL* offered to our readers is not very different from the *objective potentialities existence (OPE)* interpretation, discussed in Chapter 5.

The Wigner friend paradox in the *QL* interpretation, as is the case in *OPE*, can only be solved by postulating the existence of only one Boolean consciousness in which all observers participate, or in Leibniz's terms: there is one universal monad (God) due to which all other monads can communicate and perceive one world, but not many.

In summary, we must emphasize that according to the *QL* interpretation a *quantum object* is then to be identified with a non distributive lattice of its properties (qualities) and it is not to be supposed as *existing* in spacetime. Yes-no values are given to the elements of this lattice by a Boolean observer and they change in time according to the wave packet collapse rule. So, the lattice itself describes only some *objective potentialities*, which are actualized as *events* due to observation.[8]

What is the interpretation of *nonlocality* in a quantum system in the *QL* interpretation? To give an answer, let us consider a simple example.

Let there be a quantum system constituted of two spin 1/2 particles prepared in the singlet state, and an observer who measures the spin projections $S_z^{(1)}$ for particle 1 and $S_z^{(2)}$ for particle 2. The Hasse diagram for this situation is given in Fig.(7.2).

In this diagram, atoms A, B, C, D correspond respectively to the projectors onto the states (see Eq. (2.14))

$$\begin{aligned} \mid e_{12}\rangle &= |x^3, 1\rangle_1 \otimes |x^3, 2\rangle_2 = \begin{pmatrix} 1 \\ 0 \end{pmatrix} \otimes \begin{pmatrix} 0 \\ 1 \end{pmatrix}, \\ \mid e_{21}\rangle &= |x^3, 2\rangle_1 \otimes |x^3, 1\rangle_2 \equiv \begin{pmatrix} 0 \\ 1 \end{pmatrix} \otimes \begin{pmatrix} 1 \\ 0 \end{pmatrix}, \\ \mid e_{11}\rangle &= |x^3, 1\rangle_1 \otimes |x^3, 1\rangle_2 \equiv \begin{pmatrix} 1 \\ 0 \end{pmatrix} \otimes \begin{pmatrix} 1 \\ 0 \end{pmatrix}, \\ \mid e_{22}\rangle &= |x^3, 2\rangle_1 \otimes |x^3, 2\rangle_2 \equiv \begin{pmatrix} 0 \\ 1 \end{pmatrix} \otimes \begin{pmatrix} 0 \\ 1 \end{pmatrix}. \end{aligned} \quad (7.31)$$

78 NONLOCALITY IN QUANTUM PHYSICS

and atom E corresponds to the projector onto the state

$$| \ q-\rangle = \frac{1}{\sqrt{2}}(|e_{12}\rangle - |e_{21}\rangle)$$

$$= \frac{1}{\sqrt{2}}\left[\begin{pmatrix} 1 \\ 0 \end{pmatrix} \otimes \begin{pmatrix} 0 \\ 1 \end{pmatrix} - \begin{pmatrix} 0 \\ 1 \end{pmatrix} \otimes \begin{pmatrix} 1 \\ 0 \end{pmatrix}\right]. \qquad (7.32)$$

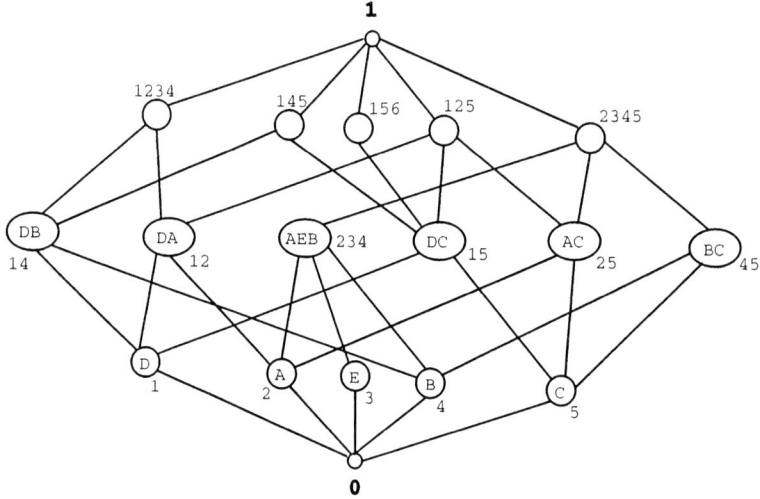

Figure 7.2. Hasse diagram for a two particles spin $1/2$ system for which $S_z^{(1)}$ and $S_z^{(2)}$ are measured.

If we take atom E away from the lattice, it becomes distributive! The non-distributivity is due to the formula

$$A \wedge (B \vee E) = A \neq (A \wedge B) \vee (A \wedge E) = 0 \vee 0 = 0. \qquad (7.33)$$

So, we cannot define a classical probability measure for the whole lattice. It is E that *makes the system nonlocal*. Here, we can again discuss as we did in Chapter 6, the problem concerning the *event-like sense*, when we have two particles separated by a space-like interval. If we know the wave function, we can give some weights to the atoms, e.g., 1 for E, $1/2$ for A, $1/2$ for B, and 0 for C and D. The element 25 means $A \vee C$, which means that the first observer detected $S_z^{(1)} = 1/2$, but this does not mean that A is true, i.e., that $S_z^{(2)} = -1/2$, despite the fact that the system was prepared in the singlet state. Here, the classical rules of *implication* do not apply to this non distributive system; indeed, we have $E \wedge (A \vee C) = 0 \neq A$.

$S_z^{(2)} = -1/2$ corresponds to the element 12, which is AD. If, for the second observer 12 is true, this corresponds to $A \vee D$. But, $A = (A \vee C) \wedge (A \vee D)$ as a consequence of the structure of our lattice. We need two observations $A \vee C$, $A \vee D$

in order to know A, and not only $A \vee C$. This solves the *relativistic paradox* for the two particles if we try to describe them in terms of events.

We have already discussed that the *birth of time* is a necessity for the observer if he desires to make his internal logic adequate for the logic of phenomena that are outside his mind. But in so doing, a *Boolean consciousness* duplicates the lattice. It becomes necessary for it to construct a new Hilbert space with *superselection rules* due to time. For t_1, we have the Hilbert space $\mathcal{H}_{t_1}$, where atom E is fixed and this corresponds, e.g., to the preparation of the state of the system. In some other Hilbert space $\mathcal{H}_{t_2}$, the Boolean consciousness fixes some weight-conditioned probabilities A, B, C, D, defined by E, e.g., 1/2, 1/2, 0, 0. So, by using time and motion in time, a Boolean observer can *imitate* a non Boolean lattice, observing at t_1 one element and at t_2 some other element of it.

It is also easy to see here, how *quantum indeterminism* appears. From Fig.(7.2) we see that the elements $1, 2, 3, 4, 5$ are incompatible. So, for a Boolean mind, if 3 is true, then $1, 2, 4, 5$, are false. But for our non distributive lattice $1 \vee 2 \vee 4 \vee 5$ is true.

The *escape* for the logic of the observer is found in the proposition, that at time t_1, 3 is true, at t_2, $1 \vee 2 \vee 4 \vee 5$ *becomes true*, and in such in a totally unknown way one of the 1,2,4,5 *becomes true*!

In section 2.4, we saw that for noncommuting operators, we can put into correspondence some commuting *operators* of frequencies, defined for an actual infinite system of copies of the same quantum system. Understanding this infinite system as an infinite number of copies at different moments of time, we see the possibility due to time of Boolenization of the non distributive lattice corresponding to the quantum system.

If the necessity of taking an *infinite* number of moments in time means some special property of time like pure duration (i.e, a finite interval, but not constituted of separate moments!) is still an open question.

Notes

1. See Bourbaki[14] for the general notion of algebraic structure.
2. At least, in the awareness state.
3. The fundamental concept of *state of a physical system* will be defined below.
4. It is a good exercise to verify that if $a \leq b'$ then $a \leftrightarrow b$.
5. See section B.6 of Appendix B.
6. It is necessary to say that many authors consider QL simply as some part of the mathematical formalism of quantum theory, without the claim of any interpretation, see, e.g., Jauch[13], Bub[19], and a good review of the discussions in Jammer[1]. We remark also that QL is related with the so-called algebraic approach to quantum field theory, first proposed by Jordan, von Neumann and Wigner[20] and developed by Segal[21], Haag and Kastler[22], and others.
7. Of course, in our interpretation of QL, Minkowski spacetime is an invention of the human mind and nothing can a priori guarantee that it is the best concept to accommodate the relationships between the properties of the phenomena that are outside the human mind.

8. It is known that in altered states of consciousness (meditation), the flow of time seems to stop[24] (the *eternal present* state). An observer in profound meditation perceives very *strange* events, which in general do not make sense if he tries to relate his experiences when in a state of awareness for someone also in a state of awareness in which Boolean logic is in operation. Let us emphasize here that the meditation experience although subjective, is universal in the sense that learning how to obtain altered states of consciousness is open to anyone. Some facts from psychiatry may become more intelligible if the change from Boolean to quantum logic is taken into account[25].

References

[1] M. Jammer, *The Philosophy of Quantum Mechanics* (J. Wiley & Sons, New York, 1974).

[2] G. Birkhoff and J. von Neumann, The logic of quantum mechanics, *Ann. Math.* **37**, 823–843 (1936).

[3] C. Piron, Axiomatique quantique, *Helv. Phys. Acta* **37**(4-5), 439–468 (1964).

[4] J. M. Jauch and C. Piron, On the structure of quantal proposition systems, *Helv. Phys. Acta.* **42**(6), 842–848 (1969).

[5] D. Finkelstein, The logic of quantum mechanics, *Trans. New York Acad. Sci.* **25**, 621–627, (1969).

[6] D. Finkelstein and S. R. Finkelstein, Computational complementarity, *Int. J. Theor. Phys.* **22**(8), 753–779 (1983).

[7] A. A. Grib and R. R. Zapatrin, Automata simulating quantum logics, *Int. J. Theor. Phys.* **29**(2), 113–123 (1990);

[8] A. A. Grib and R. R. Zapatrin, Topology lattice as quantum logic, *Int. J. Theor. Phys.* **31**(7), 1093–1101 (1992).

[9] A. A. Grib and R. R. Zapatrin, Topologimeter and the problem of physical interpretation of topology lattice, *Int. J. Theor. Phys.* **35**(3), 593–604 (1996).

[10] R. R. Zapatrin, Binary quantum logic and generating semigroups, *Int. J. Theor. Phys.* **28**(11), 1323–1332 (1989);

[11] R. R. Zapatrin, Binary logic is rich enough, *Int. J. Theor. Phys.* **31**(2), 211–219 (1992).

[12] H. Putnam, Lógica, in: *Enciclopédia Einaudi* **13**, (Imprensa Nacional-Casa da Moeda, Lisboa, 1989), pp. 11–195.

[13] J. M. Jauch, *Foundations of Quantum Mechanics* (Addison-Wesley Publishing Co., Reading, MA, 1968).

[14] N. Bourbaki, *Théorie des Ensembles* (Hermann, Paris, 1957).

[15] G. Birkhoff, *Lattice Theory* (third edition), AMS Colloc. Publ., vol. **25** (AMS, Providence, Rhode Island, 1993).

[16] C. Piron, *Axiomatique Quantique*, Ph.D. thesis, Univ. Lausanne, 1993.

[17] H. Putnam, *Boston Studies in the Phil. Sci.* **5**, 216–241 (1969).

[18] V. S. Varadarajan, *Geometry of Quantum Theory*, vols. 1 and 2 (Van Nostrand Reinhold Co., New York, 1970).

[19] J. Bub, On the Completeness of Quantum Mechanics, in: *Contemporary Research in the Foundations of Quantum Mechanics*, edited by C. A. Hooker (Reidel, Dordrecht, 1973), pp.1–65.

[20] P. Jordan, J. von Neumann, and E. P. Wigner, On an algebraic generalization of the quantum mechanics formalism, *Ann. Math.* **35**, 29–64 (1934).

[21] I. E. Segal, Postulates for general quantum mechanics, *Ann. Math.* **48**, 930–948 (1947).

[22] R. Haag and D. Kastler, An algebraic approach to quantum field theory, *J. Math. Phys.* **5**(7), 848–861 (1964).

[23] A. A. Grib, Quantum logical interpretation of quantum mechanics—the role of time, *Int. J. Theor. Phys.* **32**(12), 2389–2400 (1993).

[24] K. Ramakrishna Rao, Meditation and Mind/Matter Interface, in: *The Interrelationship Between Mind and Matter*, edited by B. Rubik (The Center for Frontier Sciences, Temple University, Philadelphia, PA, 1992), PP.223-.

[25] A. A. Grib, in: *Philosophical Problems — Quantum Mechanics and Relativity Theory* (in memory of V. A. Fock) (in Russian), edited by P. P. Pavinsky (Izdatelstvo Leningradskogo Universiteta, 1980), pp. 130–144.

8
MACROSCOPIC REALIZATIONS OF QL

If the quantum theory for the microworld is just a realization of QL, we can ask: are there other examples, macroscopic ones, where the same logic can be realized. Examples of this kind have been given by Grib and Zapatrin[1--4] which showed Bell's inequalities invalidated. In what follows, we discuss in details some examples of macroscopic realizations of QL. A very important feature of the examples to be analyzed is that each one is described by the same mathematical formalism as the one appropriate for quantum systems: Hilbert space, noncommuting self-conjugate operators for observables, Heisenberg uncertainty relations *without* Planck's constant. All this, for situations that seem completely different from the ones in the microworld.

8.1 LATTICES AND GRAPHS

In order to present such examples we must use the *correspondence*, noted by Finkelstein and Finkelstein[5], between quantum logical lattices and graphs of some classical system.

We start by given the mathematical definition of a graph and their properties.

By definition a *graph* is any set of ordered pairs of the elements of a set E, i.e. :

$$\mathfrak{G} \text{ is a graph} \Leftrightarrow (\forall z \in \mathfrak{G} \Rightarrow (\exists x), (\exists y) \in E | z = (x, y)). \tag{8.1}$$

The *domain* of a graph (or its first projection) is the set denoted

$$Dom(\mathfrak{G}) = \{x \in E | \exists y \in E | (x, y) \in \mathfrak{G}\}. \tag{8.2}$$

The image of a graph (or second projection, or the set of the values) is the set denoted

$$Im(\mathfrak{G}) = \{y \in E | \exists x \in E | (x,y) \in \mathfrak{G}\}. \tag{8.3}$$

In the case where $Dom(\mathfrak{G})$ and $Im(\mathfrak{G})$ are finite sets we can represent the graph $\mathfrak{G}$ by a *sagittal diagram* using oriented arcs, through the following convention: an element $x \in Dom(\mathfrak{G})$ is connected by an arc to an element $y \in Im(\mathfrak{G})$, if and only if $(x,y) \in \mathfrak{G}$. The points of the sagittal diagram are called the *vertices* of the graph. Two vertices are said to be *adjacent*, if they are connected with an arc. Each vertex is by definition adjacent to itself. The set of vertices will be denoted $V(\mathfrak{G})$ and the set of arcs by $Arc(\mathfrak{G})$. A graph such that if $(x,y) \in \mathfrak{G}$, then $(y,x) \in \mathfrak{G}$ is said to be nonoriented.

As an example consider the sagittal diagram for the graph

$$\mathfrak{G} = \{(1,1),(1,2),(2,3),(3,2),(4,3)\}. \tag{8.4}$$

Then,

$$Dom(\mathfrak{G}) = \{1,2,3,4\}, \ Im(\mathfrak{G}) = \{1,2,3\}. \tag{8.5}$$

The sagittal diagram is represented in Fig.(8.1). Note the usual convention used in the diagram on the right of omitting the closed arc corresponding to the connection between the element 1 with itself, and the omission of the arrows between the points 2 and 3.

Figure 8.1. Equivalent sagittal diagrams for the graph of Eq. (8.4).

Consider now, the Hasse diagram for the lattice of propositions of a spin 1/2 particle corresponding to the measurements (not simultaneously, of course) of the non-commuting spin projections S_z, S_x.

On the right hand side of Fig.(8.2), we have the graph $\mathfrak{G}_U$ of a classical system.

$$\begin{aligned}\mathfrak{G}_U =\ & \{(1,1),(1,2),(1,4),(2,1),(2,2),(2,3),(3,2),(3,3),(3,4),(4,1),\\ & (4,3),(4,4)\}\end{aligned} \tag{8.6}$$

To fix the ideas, let our classical system be a university with four states (economical ones) indexed by $n = 1, 2, 3, 4$, which is supposed to have the following *properties*:

(i) It answers *yes* on the question asked by the observer: are you in the state n? — not only if it is in the state n, but also if it is in the states $(n - 1)[\mod 4]$ or $(n + 1)[\mod 4]$.

(ii) It can *change* its state by one step, *reacting* on the question asked by the observer, if and only if the states are connected with an arc of the graph.

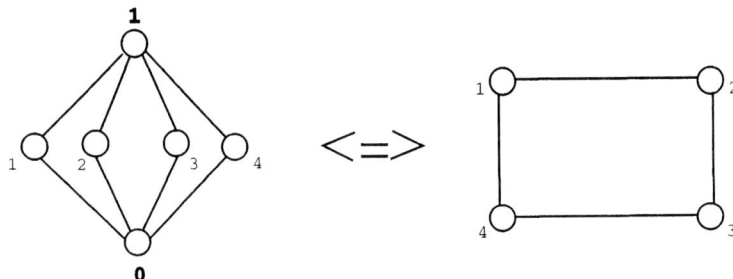

Figure 8.2. Hasse diagram (left) for the lattice of particular propositions concerning the spin components for a spin $1/2$ particle and the graph (right) of a classical system simulating the quantum system.

Let there be some observer (commission from the federal government) that wants to define what state our system is in, and let us imagine that the commission is clever enough to guess the above properties of the investigated system. Then, the strategy to arrive at some conclusion is to use the following *negative* logic, and say: the system is in state 2, if with the question: are you in state 4? it will get a negative answer. In this way, obtaining negative answers on complementary questions, the commission can learn the *true* states of the ensembles of similar systems, described by the graph in Fig.(8.2).

But, the negative logic leads to situations, such that there are no questions, for which negative answers means "1 or 2","2 or 3", etc. "1 or 2" is equivalent to **1**-any state! It is impossible to see the difference between the disjunctions $1 \vee 2, 2 \vee 3, 3 \vee 4, 1 \vee 4$, and **1**-always true (any state). This leads to the result that the logic of the observer is described by the Hasse diagram on the left of Fig.(8.2). The reason for the non-distributivity here is the special property of disjunction ; $1 \vee 2$ is true, if 1 is true, $1 \vee 2$ is true, if 2 is true, but not only if: $1 \vee 2$ can be true if both 1, 2 are false, but if 3 is true!

We could substitute the university model with one of automata (or computers) where quantum logic can arise due to our general scheme.

8.1.1 GRIB AND ZAPATRIN RULE FOR CONSTRUCTION OF THE LATTICE CORRESPONDING TO A GIVEN GRAPH AND VICE-VERSA

Grib and Zapatrin[1] found simple algorithms that permit us to construct a lattice corresponding to a given graph (it is a simplification of the method proposed by Finkelstein and Finkelstein[5]), and the opposite, how to construct a graph corresponding to a given lattice. The algorithms make constructing classical systems *simulating* some known quantum objects possible, and can be used for purposes of analogous computations in quantum physics.

To understand the idea behind the construction of the algorithms, let us first consider a classical system which can have only N distinct states. It can be represented as a disconnected graph, as in Fig.(8.3).

$$\bullet\ \bullet\ ......\ \bullet$$
$$1\ 2......N$$

Figure 8.3. Graph for a classical system with N distinct states.

The set $S = \{1, 2....N\}$ is a phase space for the system. The Boolean lattice 2^N of the subsets of S is the lattice of propositions of the classical system, which can be constructed as follows;

Ground floor: **0**.

First story: consists of the singletons $1, 2, ...N$.

Second story: consists of two-elements sets $\{1, 2\} \equiv 12, \{1, 3\} \equiv 13, \ldots \{N-1, N\}$

...

N-th story: the set S itself.

The lattice is self-dual in the sense that if we turn it over and switch each element by for complement in relation to S we again obtain the same lattice 2^N.

To imitate quantum systems classically, we use the concept of a normalized automaton.

By definition, a *normalized automaton* $\mathfrak{A}$ is an automaton defined by a nonoriented graph $\mathfrak{G}$, satisfying the conditions:

(i) The set of input symbols and the set of interior states of $\mathfrak{A}$ coincide with $V(\mathfrak{G})$, the set of vertices of $\mathfrak{G}$.

(ii) The transition function (i.e., the rule of operation of the automaton) is defined in the following way. If the automaton is initially in some state i and the input symbol is j, then if the vertices are adjacent, the new state will be j, and if they are not connected, the automaton stops its working.

$$(i,j) = \begin{cases} j & \text{if } (i,j) \in Arc(\mathfrak{G}) \\ \text{stop} & \text{otherwise} \end{cases} \quad (8.7)$$

The graph is reflexive.

Now, to each nonoriented graph we can correspond a lattice of its observed properties (lattice of propositions), using the following general rule (algorithm). First we construct with the rules (i-iii) below, the so called *union lattice*, and then use rule (iv) to obtain the *lattice of propositions* of the normalized automaton.

(i) At the foundation of the union lattice we put the *empty set-* **0**.

(ii) The first story (or level) of the union lattice consists of *stars* of the vertices of ₲. The star $[i]$ of the vertex i- is the set of all vertices adjacent to i.

(iii) Construct all set theoretical unions of stars. Then construct the new levels of the union lattice according to the number of elements of the unions obtained above. The top level is then the set $V(₲)$ of vertices of the graph ₲.

(iv) Invert the lattice obtained, changing each element by its set theoretical complement.

The result of (i-iv) is the lattice of the propositions corresponding to the given graph defining the normalized automaton.

As one example of the application of the above rule, consider the graph on the left side of Fig.(8.4).

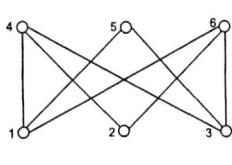
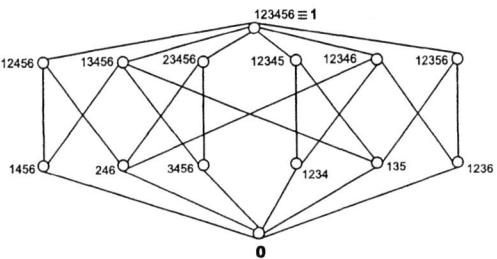

Figure 8.4.   A graph (left) and corresponding lattice (right).

Using the rule, we construct the lattice shown on the right side of Fig.(8.4). The first three steps are:

(i) at the bottom, we put-0,

(ii) the first level is: $[1] = 1456, [2] = 246, [3] = 3456, [4] = 1234, [5] = 135, [6] = 1236$,

(iii) $[1] \cup [2] = 12456, [1] \cup [3] = 13456, [1] \cup [4] = 123456, [1] \cup [5] = 123456, [1] \cup [6] = 123456, [2] \cup [3] = 23456, [2] \cup [4] = [2] \cup [5] = [2] \cup [6] = 123456, [3] \cup [4] = [3] \cup [5] = [3] \cup [6] = 123456, [4] \cup [5] = 12345, [4] \cup [6] = 12346, [5] \cup [6] = 12356$.

With (i-iii) we get the union lattice on the left side of Fig.(8.5). Inverting the union lattice and taking the set theoretical complements for the elements we finally obtain the lattice of propositions, which is on the right side of Fig.(8.5).

The lattice of propositions obtained is not distributive, but it is modular, e.g., we have:

$$1 \wedge (2 \vee 6) \neq (1 \wedge 2) \vee (1 \wedge 6)$$
$$1 \wedge (2 \vee 3) = (1 \wedge 2) \vee (1 \wedge 3). \qquad (8.8)$$

88 NONLOCALITY IN QUANTUM PHYSICS

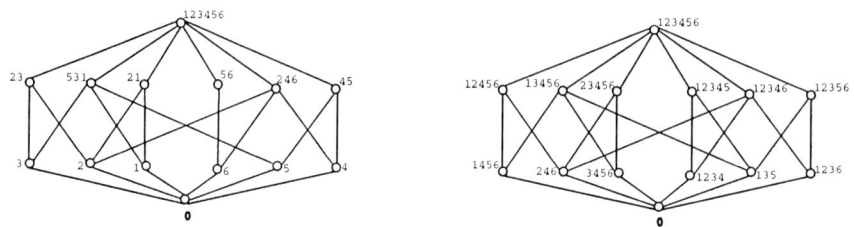

Figure 8.5. (Left) Steps for constructing the lattice of Fig.(8.4); (Right) The constructed lattice.

It is very important to observe that we can obtain for some nonoriented automata lattices of propositions that are not modular. The simplest example is given by the normalized automaton with the graph given by Fig(8.6).

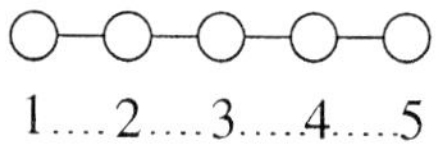

○—○—○—○—○
1....2....3....4....5

Figure 8.6. A simple graph of a normalized automaton.

To the above graph our rule gives the nonmodular lattice whose Hasse diagram is given in Fig.(8.7).

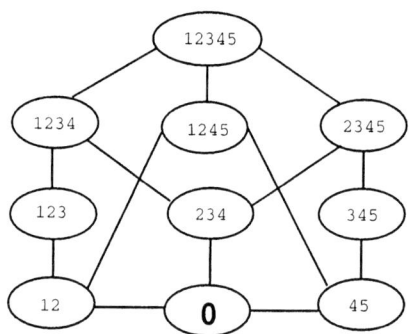

Figure 8.7. Hasse diagram for the lattice corresponding to the graph of Figure 8.6.

Let $\mathfrak{G}_1$ and $\mathfrak{G}_2$ be two arbitrary graphs corresponding to two nonoriented automata and let $\mathfrak{L}(\mathfrak{G}_1)$ and $\mathfrak{L}(\mathfrak{G}_2)$ be the corresponding lattices of propositions. Then, we have:

Theorem 1. If $\mathfrak{G}$ is the graph obtained by placing together two finite graphs $\mathfrak{G}_1$ and $\mathfrak{G}_2$, we write

$$\mathfrak{G} = \mathfrak{G}_1 + \mathfrak{G}_2 \tag{8.9}$$

then

$$\mathcal{L}(\mathfrak{G}) = \mathcal{L}(\mathfrak{G}_1) \times \mathcal{L}(\mathfrak{G}_2) \qquad (8.10)$$

where × means the union lattice product defined in Appendix A.

Theorem 2. If $\mathfrak{G} = \mathfrak{G}_1 \circ \mathfrak{G}_2$ is a *bunch* of $\mathfrak{G}_1$ and $\mathfrak{G}_2$, i.e., each vertex of $\mathfrak{G}_1$ is connected by an arc with each vertex of $\mathfrak{G}_2$, then

$$\mathcal{L}(\mathfrak{G}) = \mathcal{L}(\mathfrak{G}_1) + \mathcal{L}(\mathfrak{G}_2) \qquad (8.11)$$

where + means in the above equation, the so-called *horizontal sum*. This means that:

(i) $\mathcal{L}(\mathfrak{G}_1) + \mathcal{L}(\mathfrak{G}_2)$ is the set of all elements in $\mathcal{L}(\mathfrak{G}_1)$ or $\mathcal{L}(\mathfrak{G}_2)$ considered as disjoint. $\leq$ keeps its meaning in $\mathcal{L}(\mathfrak{G}_1)$ and in $\mathcal{L}(\mathfrak{G}_2)$, but for $\forall x \in \mathcal{L}(\mathfrak{G}_1), \forall y \in \mathcal{L}(\mathfrak{G}_2)$, neither $y \leq x$ nor $x \leq y$.

(ii) The greatest and least elements of both lattices are identified, i.e.,

$$\mathbf{1}_{\mathcal{L}(\mathfrak{G}_1)} \equiv \mathbf{1}_{\mathcal{L}(\mathfrak{G}_2)} = \mathbf{1}_{\mathcal{L}(\mathfrak{G})} = \mathbf{1}, \quad \mathbf{0}_{\mathcal{L}(\mathfrak{G}_1)} \equiv \mathbf{0}_{\mathcal{L}(\mathfrak{G}_2)} = \mathbf{0}_{\mathcal{L}(\mathfrak{G})} = \mathbf{0} \qquad (8.12)$$

The rule for constructing the graph of a normalized automaton corresponding to the Hasse diagram of a lattice is the following: to each atom of the lattice corresponds a vertex of the graph. Then connect by lines all nonorthogonal states ($a \perp b \Leftrightarrow a \leq b'$).

As an example, consider an electron, which is a spin 1/2 particle, with three noncommuting spin projections, let us say, S_x, S_y, S_z. The Hasse diagram for this quantum system is on the left side of Fig.(8.8) and the corresponding graph is on the right side of the same figure.

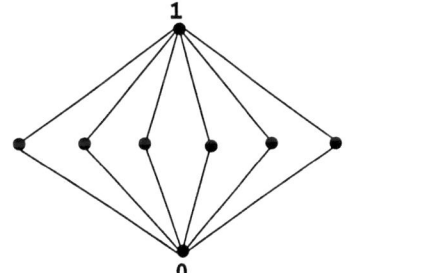
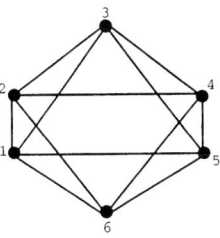

Figure 8.8. Left: Hasse diagram for the lattice of propositions concerning the spin components of a spin 1/2 particle. Right: Graph for a classical system simulating the quantum system.

If we have a *classical system* with the graph as in Fig.(8.8), it is possible to obtain an analog for it of the Heisenberg uncertainty relations for the spin projections. An ensemble of such classical systems will be described not by a *probability distribution* (if the states are stochastic!), but by a *wave function*. The wave function is just some state for the spin 1/2 electron, which as we know is a vector $|\Psi\rangle \in \mathcal{H} \approx \mathcal{C}^2$.

For a spin 1/2 electron there are the following *yes-no* questions: $S_x = 1/2?$, $S_y = 1/2?$, $S_z = 1/2?$, $S_x = -1/2?$, $S_y = -1/2$?, $S_z = -1/2$? to which correspond projectors $\mathbf{P}_\alpha$ ($\alpha = 1, 2, ...6$).

To the wave function $|\Psi\rangle$ we correspond the weights $\{a_\alpha\}$, where $a_\alpha = |\mathbf{P}_\alpha|\Psi\rangle|^2$, such that to a vertex α of the graph there corresponds a_α. It is easy to obtain,

$$a_1 + a_4 = a_2 + a_5 = a_3 + a_6 = 1$$
$$\sum_{\alpha=1}^{n} a_\alpha^2 = 2. \tag{8.13}$$

So, our weights do not have the meaning of the standard frequencies of events as in the usual probability theory.

Now, let us introduce *observables* on the graph. Consider an observable A and the following 2×2 Hermitian matrix

$$\mathbf{A} = \begin{bmatrix} a & b + ic \\ b - ic & d \end{bmatrix} \tag{8.14}$$

which is the matrix representation of A in the matrix representation of spin space, and where $a, b, c, d \in \mathcal{R}$. Now, put $\mathbf{A}$ in correspondence with

$$A_\alpha = Tr(\mathbf{P}_\alpha \mathbf{A}), \quad \alpha = 1, 2, ...6 \tag{8.15}$$

such that,

$$2A_1 = a + 2b + d, \ A_3 = a, \ 2A_5 = a + 2c = d$$
$$2A_2 = a - 2c + d, \ 2A_4 = a - 2b + d, \ A_6 = d. \tag{8.16}$$

Then for the expectation value of $\mathbf{A}$ we have,

$$\overline{A} = \langle\Psi|\mathbf{A}|\Psi\rangle = \sum_{\alpha=1}^{6} A_\alpha a_\alpha - Tr\mathbf{A} \tag{8.17}$$
$$= A_1 + A_4 = A_2 + A_5 = A_3 + A_6.$$

Then, each observable A is defined by six numbers A_α in the graph. For example, if $\mathbf{A} = \mathbf{S}_x$, then $A_1 = 1/2$, $A_4 = -1/2$, all others $A_\alpha = 0$. If $\mathbf{B} = \mathbf{S}_y$, then $B_2 = 1/2$, $B_5 = -1/2$, all others $B_\alpha = 0$.

Given an arbitrary $|\Psi\rangle$, defined by six weights a_α, we can obtain for the dispersions $\mathcal{D}_\Psi \mathbf{A}$ and $\mathcal{D}_\Psi \mathbf{B}$ a *Heisenberg uncertainty relation*, without any Planck's constant. Indeed,

$$(\mathcal{D}_\Psi \mathbf{A})(\mathcal{D}_\Psi \mathbf{B}) \geq 1/4(\overline{S}_z)^2, \ \overline{S}_z = \langle\Psi|\mathbf{S}_z|\Psi\rangle \tag{8.18}$$

which looks like,

$$a_1 a_4 a_2 a_5 \geq \frac{1}{16}(a_3 - a_6)^2. \tag{8.19}$$

8.2 BREAKDOWN OF BELL'S INEQUALITIES

Having obtained the rule making the construction of classical systems that satisfy *QL* and thus have wave functions possible, we can ask about the *breakdown* of Bell's inequalities for these systems. We recall that Bell's inequalities in d'Espagnat form (Chapter 4) have been obtained using the distributivity law, which is valid for usual sets, and then we expect that it is an easy task to obtain the breakdown of Bell's inequalities for our systems.

Suppose we have a two-particle system, such that each one of the particles is described by a (different) quantum logic lattice. Each particle *is* as some *beable*, but considering the whole we need to use non distributive *QL*. Can we say that the whole is a *beable* together with the parts as *beables*?

If the answer is *yes*, then we get Bell's inequalities. But, to obtain the *right* answer to our question, it is necessary to recall that, as explained before, when a Boolean observer examines a system satisfying *QL*, he gives truth values that are different at different instants of time. For our *classical* example represented by the university departments, there exists an *interaction* between the observer using *negative* logic and the system, each time that a question is asked. Then it is clear that the truth values *arise* from this interaction.

For a two-particle system, it is necessary to add new elements corresponding to the *permutation* operator to the lattice. This operator, as we know, does not commute with local operators and as a consequence, as we saw before, even if only the spin projections $S_z^{(1)}$ and $S_z^{(2)}$ are measured for the particles 1 and 2, the lattice becomes non distributive, while it was distributive before taking the permutation operator into account. So, we obtain *complementarity* between the whole and its parts. There are *leaps* for the *truth values* if some Boolean observer makes measurements.

Now, following[4], consider *two classical systems*, such that to each one of them corresponds a graph with four vertices as on the right hand side of Fig.(8.2). We already know that to a graph of this kind there corresponds a non Boolean lattice that describes a spin 1/2 system for which, e.g., S_x and S_z are measured. How to construct the *graph* for the two classical systems?

Let $\mathcal{H}^1$ and $\mathcal{H}^2$ be the Hilbert spaces for each one of the particles, and let i be some question relative to the first particle and k be some question relative to the second particle. Then, to the pair of questions ik, in quantum theory there corresponds a vector

$$|ik\rangle \equiv |i\rangle \otimes |k\rangle \in \mathcal{H}^1 \otimes \mathcal{H}^2. \tag{8.20}$$

To construct the graph $\mathfrak{G}$ corresponding to the lattice of propositions of the system, we use the rule that the vertices ik of the graph are connected with the vertices $i'k'$ if and only if,

$$\langle ik|i'k'\rangle \neq 0 \tag{8.21}$$

which is possible only if $\langle i|i'\rangle \neq 0$, $\langle k|k'\rangle \neq 0$, so that i, i' are connected by an arc in the graph of the first particle and k, k' are connected by an arc in the graph of the second particle. So, the rule to construct $\mathfrak{G}$ is the following. Let $\mathfrak{G}_1$ and $\mathfrak{G}_2$ be the graphs respectively for the objects 1 and 2. Then, the set $V(\mathfrak{G})$ of the vertices of the graph $\mathfrak{G}$ is the Cartesian product of the sets $V(\mathfrak{G}_1)$ and $V(\mathfrak{G}_2)$ of the vertices of the graphs $\mathfrak{G}_1$ and $\mathfrak{G}_2$, i.e.,

$$V(\mathfrak{G}) = V(\mathfrak{G}_1) \times V(\mathfrak{G}_2). \tag{8.22}$$

The set of arcs $Arc(\mathfrak{G})$ of the graph of $\mathfrak{G}$ is also the Cartesian product of the set of arcs of the corresponding graphs.

Let us apply these rules for a system of two spin 1/2 particles. We obtain a graph $\mathfrak{G}$ with $4 \times 4 = 16$ vertices. It is possible to draw it as in Fig.(8.9), where the vertices with the same indices are identified.

Once we have this graph, it is possible to construct the corresponding quantum lattice using the rule given above. This lattice, going from *up* to *down* is as follows: on the top there is **1**. Then, there are 16 elements, on the next level there are 72 elements, then in the other level there are 16 elements, and finally below there is **0** on the bottom.

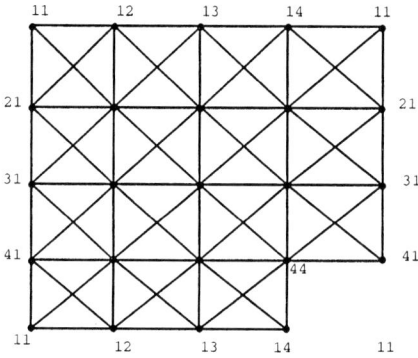

Figure 8.9. Graph for two spin 1/2 particles.

As we said above, in order to break Bell's inequalities and obtain the *whole-parts complementarity*, it is necessary to take into account the permutation operator, which has two eigenvalues ± 1. To the eigenvalue -1 there corresponds the eigenvector $|q_-\rangle \in \mathcal{H}^1 \otimes \mathcal{H}^2$, such that,

$$|q-\rangle = \frac{1}{\sqrt{2}}(|e_{12}\rangle - |e_{21}\rangle), \tag{8.23}$$

and to the eigenvalue $+1$ there corresponds its *orthocomplement*.

Let us put a new element into a given graph $\mathfrak{G}$ corresponding to the question: *are the particles in an antisymmetric state?* Then, connect this element with the other vertices of the graph $\mathfrak{G}$ using the same rule already introduced: if the corresponding subspaces are non orthonormal, then connect the vertices by an arc. Calculating the scalar products $\langle ik|q_-\rangle$ we obtain that $|q-\rangle$, is orthogonal to 11, 22, 33, 44. So, connect $|q_-\rangle$ with arcs with the vertices different from these ones. Concerning $|q_+\rangle$ we have $\langle ik|q_+\rangle \neq 0$ for all ik and then it must be connected with all the other vertices.

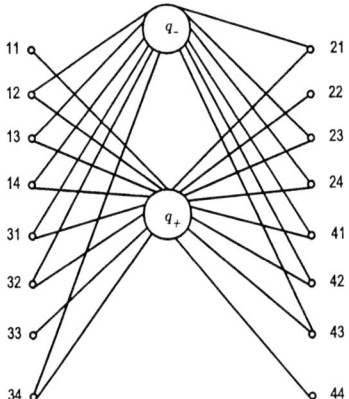

Figure 8.10. Graph for two spin 1/2 particles including antisymmetrization.

In this way, we obtain a new graph, $\widetilde{\mathfrak{G}}$ which is shown in Fig.(8.10).

Let us now construct the quantum lattice corresponding to the graph $\widetilde{\mathfrak{G}}$. The maximal element **1** is the set of all vertices $V(\widetilde{\mathfrak{G}})$, which has 18 elements. The next level has 18 elements. It is important to pay attention to the orthogonal elements,

$$(q_-)^\perp = (11, 22, 33, 44, q_+), \quad (q_+)^\perp = q_- \quad (8.24)$$

and take notice that q_+ is not atomic since,

$$(q_+)^{\perp\perp} = (11, 22, 33, 44, q_+) \neq q_+. \quad (8.25)$$

So, the lattice $\mathfrak{L}(\widetilde{\mathfrak{G}})$ has two new elements compared with the lattice $\mathfrak{L}(\mathfrak{G})$, respectively the atom q_- and the element $(q_+) = (q_+, 11, 22, 33, 44)$. These elements are connected with the following elements of the lattice: from (q_+) there are lines that go down to $(11, 22)$, $(11, 33)$, $(11, 44)$, $(22, 44)$ and $(33, 44)$; from q_- there are lines going up to $(34, 43)$, $(31, 13)$, $(32, 23)$, $(14, 41)$, $(24, 42)$ and $(12, 21)$. In this way, the lattice $\mathfrak{L}(\widetilde{\mathfrak{G}})$ is completely described.

The state $|\Psi\rangle$ of a two particles system is described by a set of weights $\{d_{mn}\}$. If the state is factorizable, i.e., it is of the form given by Eq. (8.20), then $d_{mn} = |\langle mn|\Psi\rangle|^2$ is the product of weights

$$d_{mn} = f_m g_n. \quad (8.26)$$

The probability of getting a positive answer to the question: is the system in the state $|ik\rangle$, which is defined by the weights $a^i b^k$ for the vertices ik? — is given by the formula

$$p_{ab,d} = a^i b^k T^{mn}_{ik} d_{mn} + 5/4$$

$$T^{mn}_{ik} = \begin{cases} 1 & i = k, \ m = n \\ 0 & (ik) \perp (mn) \\ \frac{-1}{4} & \text{in all other cases} \end{cases} \quad (8.27)$$

Consider now the antisymmetric state,

$$|q-\rangle = \frac{1}{\sqrt{2}} \left[\begin{pmatrix} 1 \\ 0 \end{pmatrix} \otimes \begin{pmatrix} 0 \\ 1 \end{pmatrix} - \begin{pmatrix} 0 \\ 1 \end{pmatrix} \otimes \begin{pmatrix} 1 \\ 0 \end{pmatrix} \right]. \quad (8.28)$$

To it corresponds the matrix of weights $d_{mn} = |\langle mn|q-\rangle|^2$ for the graph given by

$$d_{mn} = \begin{bmatrix} 0 & \frac{1}{4} & \frac{1}{2} & \frac{1}{4} \\ \frac{1}{4} & 0 & \frac{1}{4} & \frac{1}{2} \\ \frac{1}{4} & \frac{1}{4} & 0 & \frac{1}{4} \\ \frac{1}{2} & \frac{1}{4} & \frac{1}{4} & 0 \end{bmatrix} \quad (8.29)$$

Let X, Y, Z, be some elementary questions that can be asked about a nonoriented automaton (which is a classical system) with graph ⑤ (Fig.8.8). Let $\overline{X}, \overline{Y}, \overline{Z}$, be the negation of these questions. Consider the system of two identical objects with the property: if we ask the same question (choosing one of X, Y, Z) for the first and second objects, we always get one positive and one negative answer. Then, e.g., asking X for the first object and Z for the second, we become aware of the values X and Z for each object. We can derive Bell's inequalities,

$$P(X_1 Y_1) + P(X_1 Z_1) \geq P(Y_1 Z_1) \quad (8.30)$$

or,

$$P(X_1 \overline{Y}_2) + P(X_1 \overline{Z}_2) \geq P(Y_1 \overline{Z}_2) \quad (8.31)$$

where, in obvious notation, $P(X_1 \overline{Y}_2)$ denotes the probability for a positive answer regarding question X for the first object and negative answer for question Y for the second object. Now, take X as "is $S_x = 1$?", Y as "is $S_\alpha = +1$?", where S_α is the spin projection forming an angle α with the z-axis in the xz plane, and Z as "is $S_z = -1/2$?". To these questions on graph ⑤, there correspond the following weights:

$$\begin{array}{llllll} X: & X_1 = \frac{1}{2} & X_2 = 1 & X_3 = \frac{1}{2} & X_4 = 0 & \\ Y: & Y_1 = \frac{1}{2}\beta_+ & Y_2 = \frac{1}{2}\gamma_+ & Y_3 = \frac{1}{2}\beta_- & Y_4 = \frac{1}{2}\gamma_- & (8.32) \\ Z: & Z_1 = 1 & Z_2 = 1 & Z_3 = \frac{1}{2} & Z_4 = \frac{1}{2} & \end{array}$$

where $\beta_\pm = (1 \pm \sin\alpha)$, $\gamma_\pm = (1 \pm \cos\alpha)$.

For the questions $\overline{Y}, \overline{Z}$ the weights are: $\overline{Y}_i = 1 - Y_i$, $\overline{Z}_i = 1 - Z_i$.

Now, suppose that the system (which is really described by the graph ⑤) is prepared in the state described by the weights d_{mn} such that (q_-) is true. Then using our formula for the probabilities, we get

$$P(X_1 \overline{Y}_2) = 1 - \frac{\cos\alpha}{2}, \quad P(\overline{X}_1 \overline{Z}_2) = 1/2, \quad P(Y_1 \overline{Z}_2) = 1 + \frac{\sin\alpha}{2} \quad (8.33)$$

Putting these values in Bell's inequalities we get

$$1 - \cos\alpha \geq \sin\alpha \qquad (8.34)$$

and we see that the inequality does not hold for some values of α, e.g., $\alpha = 1/4$.

We conclude that the use of QL for some special classical systems can lead to a *breakdown* of Bell's inequalities.

8.3 THE PROBLEM OF NONLOCALITY

For some classical systems composed of parts[4], even supposing the absolute validity of special relativity, it is easy to see that we can have the *breakdown* of Bell's inequalities for some *special* observables that have nothing to do with local observables, when these observables are measured for a situation where the parts are separated by a space-like interval.

Now, classical systems with *special reactions* to the measurement of some observables cannot manifest nonlocality for these observables, because, as explained before, nondistributivity is obtained due to the special property of the disjunction "*or*". If there is some hidden variable, which becomes known to the observer in the measurement process, in such a way that the disjunction becomes *unique*, then we turn back to the case of Boolean logic, and the validity of special relativity prohibits the possibility of any nonlocal influence from one of the parts of the system on the other.

In conclusion, we can say that what our examples show is that the *mistery* of quantum nonlocality and the impossibility of description of quantum systems as classical systems is not only due to the nondistributivity of QL, but also due to the supposition that special relativity is strictly true. It is then, necessary to have a breakdown of special relativity if we want to produce a realistic (in Einstein's sense) interpretation of quantum mechanics. One possibility is the introduction of *superluminal signals*, another is to have waves that go backwards in time (along the light cone), called *advanced potentials* in electrodynamics. If this is allowed, then there are lines connecting our element $(q-)$ in the graph that go forward and backward in time (see Cramer[6]). Unfortunately, we do not have enough experimental evidence on waves going backward in time[1] or superluminal waves[2].

Notes

1. But see in this connection the papers by Beauregard quoted in Chapter 6.
2. Concerning superluminal waves see footnote no. 3 of Chapter 4 and references [7],[8].

References

[1] A. A. Grib and R. R. Zapatrin, Automata simulating quantum logics, *Int. J. Theor. Phys.* **29**(2), 113–123 (1990).

[2] A. A. Grib and R. R. Zapatrin, Topology lattices as quantum logics, *Int. J. Theor. Phys.* **31**(7), 1093–1101 (1992).

[3] A. A. Grib, Quantum logical interpretation of quantum mechanics—the role of time, *Int. J. Theor. Phys.* **32**(12), 2389–2400 (1993).

[4] A. A. Grib and R. R. Zapatrin, Quantum logical description of two-particle systems, *Int. J. Theor. Phys.* **30**(7), 949–959 (1991).

[5] D. Finkelstein and S. R. Finkelstein, Computational complementarity, *Int. J. Theor. Phys.* **22**(8), 753—779 (1983).

[6] J. Cramer, The transactional interpretation of quantum mechanics, *Rev. Mod. Phys.* **58**(3), 647–687 (1986).

[7] P. Saari and K. Reivelt, Evidence for X-shaped propagation-invariant localized light waves, *Phys. Rev. Lett.* **79**(21), 4135–4138 (1997).

[8] E. Capelas de Oliveira and W. A. Rodrigues, Jr., Superluminal Electromagnetic waves in free space, *Ann. der Physik* **7**(7-8) 654–659 (1998).

9
HIDDEN VARIABLES THEORIES (HVT)

9.1 WHAT IS A HIDDEN VARIABLES THEORY?

The attentive reader of this book certainly noted that we introduced the quantum theory (QT) through its formal language (the rules of the game, section 2.1) and then we discussed some possible meanings of the formalism and the problems associated with it. The problems of the *wave packet reduction* and the *wholeness* of "composed" systems manifested in *EPR* like situations required for their solutions until now, interpretations of the formalism that eventually may have caused an uncomfortable feeling for some[1]. For, to make sense of the formalism, we had to give a privileged status to the *consciousness* of the observer. This consciousness in some of the interpretations is responsible for the collapse of the the wave function, in others for a split in many copies of itself (in the many worlds...) each time a measurement occurs, or even for building time and space as the result of adjusting its (unfortunately) Boolean logic to the quantum logic of *things* called quantum systems, which live in a "world" that seems to be exterior to the ones being the ideal world of mind processes.

Common to all interpretations studied so far is that the description of the possible states of a given quantum system correspond to the vectors $|\Psi\rangle \in \mathcal{H}$, where $\mathcal{H}$ is the Hilbert space of the system and also that the description is *ultimate* and *complete*. $|\Psi\rangle$ is supposed to describe the single quantum system, not an ensemble of identical systems, even if the predictions of the theory can be verified only through measuring apparatuses (that in some way "materialize" the possible potential values of the observables of the quantum system) if we have a homogeneous ensemble $\mathcal{E}_{|\Psi\rangle}$ of identical systems[2]. In this sense $|\Psi\rangle$ is subjective in the epistemological sense that it gives us the ultimate knowledge that we can have about the quantum system.

Note, that the rules of the game, do not even specify the existence of an objective spacetime where particles and fields existing as beables (of our immediate intuition) live. Now, consider, e.g., an observable O to which an Hermitian operator $\mathbf{O} \in \mathcal{H} \otimes \mathcal{H}^\dagger$ with a discrete and non degenerate spectrum corresponds. We write

$$\mathbf{O} = \sum_i o_i |\varphi_i\rangle\langle\varphi_i|. \tag{9.1}$$

The rules of the game (Chapter 2) tell us that if we try to determine the value of the observable O by measuring the members (individual systems) of the ensemble $\mathcal{E}_{|\Psi\rangle}$ we will find, in general, different values o_i, which are eigenvalues of $\mathbf{O}$ corresponding to the orthonormal eigenvectors $|\varphi_i\rangle$, that determine a basis $\{|\varphi_i\rangle\}$ for $\mathcal{H}$. The main motivation for what is known as a *hidden variables theories (HVT)* comes from the hope that eventually we can avoid the discomfort mentioned above relative to the introduction of consciousness, or of metaphysical assumptions, if we give up the assumption that $|\Psi\rangle$ is the *ultimate* and *complete* description of the individual system. If this is done then, the finding of the different values o_i simply means that the individual systems in the ensemble $\mathcal{E}_{|\Psi\rangle}$ are really in different microstates. To distinguish the microstates we say that the individual system is described by $|\Psi\rangle$ together with additional parameters, here collectively denoted λ – called *hidden variables*[3] which *completely determine* the state of the individual system[4].

The *meaning* of this statement is: If a measurement of O is completely described *if* we give the complete orthonormal set $\{|\varphi_i\rangle\}$) that appears in the spectral decomposition of $\mathbf{O}$, then if the pair $(|\Psi\rangle, \lambda)$ is known and the measurement is made it will select unambiguously as the result of the measurement the vector $|\varphi_m\rangle \in \{|\varphi_i\rangle\}$). Following Belinfante[5] we write,

$$|\Psi\rangle \rightarrow |\varphi_m\rangle, \ m = m(|\Psi\rangle, \lambda, \{|\varphi_i\rangle\}). \tag{9.2}$$

The specification of the set $\{|\varphi_i\rangle\}$) will be called *contextual dependence* and it is absolutely necessary in order to avoid paradoxes, as we shall see in details below.

The first important difference between *QT* and *HVT* is that in *HVT* to $\Psi(x) = \langle x|\Psi\rangle$ (where x denotes the variables denoting the configuration space of the system) it is given the *ontological*[5] status of physical field, even if of a very special nature if compared with the usual physical fields, like, e.g., the electromagnetic field. Then, $|\Psi\rangle \rightarrow |\varphi_m\rangle$ cannot mean wave packet collapse, since $\Psi(x)$ cannot disappear and it is necessary to have a well formulated theory of the measuring processes as resulting from physical interactions between the quantum system and the measuring apparatus. Two related questions that cannot be avoided are:

(i) Does the value of λ change during a measurement?
(ii) Can we give a method to know the precise value of λ?

The answer to the first question depends on the specific theory, but we cannot see any reason for λ to remain the same, and indeed this is what happens in the well developed de Broglie-Bohm theory[6].

The answer to the second question is that nobody until now has devised a method that permits to knowing the intrinsic value of the hidden variables λ, without changing them. But, if the values of the hidden variables are not known, in order to make predictions that can be really compared with experiments, *HVT* in general assume that λ fluctuate in a random way due to external perturbations, eventually due to the existence of a subquantum medium[5,6,8]. This is described with the introduction of a classical probability distribution for the possible values λ^α of the parameters λ. We write

$$p_\alpha = p(\lambda^\alpha). \tag{9.3}$$

The next step in theory building is to suppose that as in statistical mechanics, the distribution p_α tends to the equilibrium distribution $\bar{p}_\alpha = \lim_{t\to\infty} p_\alpha(t)$, supposed self-perpetuating very rapidly. How fast, depends on the particular theory, and here there is a hope to experimentally distinguish hidden variables theory and quantum theory (see below).

Obviously, a theory of this kind cannot predict with certainty the value of m in Eq. (9.2) that will result in the measurement of the observable O, and all that can be done in this case is to predict the probability

$$\overline{P}_k(|\Psi\rangle, \{|\varphi_i\rangle\}) = \sum_{(\alpha)}^{(|\Psi\rangle, k)} \bar{p}_\alpha \tag{9.4}$$

where $\sum_{(\alpha)}^{(|\Psi\rangle, k)}$ denotes the sum for all values of λ^α for the state vector $|\Psi\rangle$ and given value

$$k = m(|\Psi\rangle, \lambda^\alpha, \{|\varphi_i\rangle\}). \tag{9.5}$$

A hidden variable theory is said to be of the *first kind*[5] if it assumes the so-called *statistical postulate*, i.e., that

$$\overline{P}_k(|\Psi\rangle, \{|\varphi_i\rangle\}) = w_k(|\Psi\rangle) = |\langle\varphi_k|\Psi\rangle|^2. \tag{9.6}$$

Equation (9.6), of course, implies that theories of the first kind allow for deviations relative to the predictions of quantum theory only before the equilibrium distribution becomes effective. After the *relaxation time*, i.e., at the time for which the equilibrium distribution becomes effective, the results of the predictions of the first kind theories will be identical to the ones of quantum theory. Papaliolios[9] tried to verify this relaxation time effect and could not find any distribution different from $\bar{p}_\alpha$ for times after the preparation of the state of 0.24×10^{-13}s. This experiment shows that it is very hard to distinguish theories of the first kind from quantum theory and the insistence in the developing of such theories requires an ideological position. This issue is well discussed in many good books, such as, e.g., in [4,6,7].

9.1.1 COMMENTS ON HVT OF THE FIRST KIND

(i) In the theories of the first kind based on the statistical postulate, λ *are* well defined, and the time dependence of λ is explicitly given by an *equation of motion* for λ involving

$d\lambda/dt$ and this equation is such that it permits making $p_\alpha \to \bar{p}_\alpha$ plausible very fast in order not to be in disagreement with the results of the Papaliolios experiment.

(ii) These theories have well-defined rules for the calculation of the function $m(|\Psi\rangle, \lambda, \{|\varphi_i\rangle\})$ for all sets $\{|\varphi_i\rangle\}$ corresponding to the spectra of all observables. However, very important is that the best developed hidden variables theory, de Broglie-Bohm's has an unambiguous rule, only for observables that are a function only of position. For all other observables, the function m depends on the values of the hidden variables associated with the measuring device in a crucial way.

(iii) As the main motivation for the introduction of *HVT,* there has been the *hope* that given $|\Psi\rangle$ and λ for a quantum system, it would be possible to predict with certainty the result of the value o_k of the observable in a measurement. But in order for this prediction to come about, it is *necessary* that the measurement distinguish $|\varphi_k\rangle$ among the vectors of the orthonormal basis $\{|\varphi_i\rangle\}$ in Eq. (9.1) that appears in the spectrum of **O**. Is it then the case that *HVT* imply that we *can say* that the observable O has the value o_k?

Surprising as it may be, the answer is *no* (to give a proof we need to study Gleason's theorem (see section 9.4, but more or less intuitive arguments presented in the sequel, show that the result can after all be expected), because it may happen that the measurement can distinguish between the members of different orthonormal sets $\{|\varphi'_i\rangle\}$ of eigenvectors of **O** if such sets contain $|\varphi_k\rangle$[7]. Then, if we can also write the spectral decomposition of **O** as

$$\mathbf{O} = \sum_i o'_i |\varphi'_i\rangle\langle\varphi'_i| \qquad (9.7)$$

such that $|\varphi_k\rangle \in \{|\varphi'_i\rangle\}$, then in general we will have

$$m(|\Psi\rangle, \lambda, \{|\varphi_i\rangle\}) \neq m(|\Psi\rangle, \lambda, \{|\varphi'_i\rangle\}) . \qquad (9.8)$$

It follows that *HVT,* even if the complete description $(|\Psi\rangle, \lambda)$ is known can predict only what will be found in a well specified measurement that takes into account the complete set of eigenvectors of **O** that is selected by the measuring apparatus. We are tempted to say that *HVT* cannot predict what *is*. However this conclusion may be wrong on many occasions. Indeed, even in classical physics a proposition like: "the z-component of a *rotating ball* is J_z," is meaningless unless we specify what the z-direction is. This means that even in classical physics J_z is not what *is*, J_z can only have a meaning in a context. We already said that we admit that *HVT* of the first kind are *context dependent*.

(iv) The above issue becomes even more intelligible when we realize that measurements in realistic *HVT* are *physical* interactions, for here, as we have already said, we are not allowed to invoke the mind (or to have an ever splitting world, etc.) to collapse the wave packet $|\Psi\rangle \to |\varphi_m\rangle$. Then, it can be verified that in all realistic *HVT* of the first kind when we model a given measurement of some observable (that is not a function of position) by means of an interaction we always find that the value obtained is the one that the system has *after* the interaction, i.e., after the *specific measurement* is completed. This point has been clear since 1927, when de Broglie[10-12] proposed his theory of the pilot wave. In particular, the de Broglie-Bohm theory (which we discuss in the next two Chapters) has a reasonable theory of measurement[6,7], and of course is context-dependent.

(iv) Are these theories compatible with relativity theory? This question of fundamental importance will be discussed in the Chapter 11.

9.1.2 THEORIES OF THE SECOND AND ZEROTH KINDS

Theories of the second kind have been elaborated in the past with the explicit purpose of providing local explanations of the *EPR*-like experiments. As we already discussed in Chapter 4, these theories make predictions that are in flagrant contradiction in relation to the ones of quantum theory. Since the Aspect experiments[11−16] seem to show that the quantum theory results are the correct ones, these theories can be said to be out of the game.[8]

Very earlier in the history of quantum theory von Neumann[17](1932) , and after him Gleason[18], Kochen-Specker[19] and Jauch[20] claimed to have proven rigorous mathematical theorems (*non-go theorems*) showing the impossibility of constructing *HVT* that could predict results in agreement with quantum theory. These approaches are based on mathematical axiomatizations of quantum theory. Among the axioms, there are some that are used in the proofs of the non-go theorems, but that viable *HVT* avoid trivially. Since these theorems, although mathematically correct, are based on axioms that do not need to be fulfilled by the physical systems in the real world, Belinfante[5] suggested to call the *no* hidden variables theories of these authors, *hidden variables theories of the zeroth kind*. This does not mean that these *non-go theorems* are not important. On the contrary, they are very important since they tell us what we *cannot* expect of realistic *HVT*.

We discuss such theorems below, but before, we comment that philosophical reasons apart, there is a more *modest reason* why we think that it is useful to know specific *HVT* and particularly how they apply to some quantum systems. Indeed, many models show how we can simulate the properties of some quantum micro-objects, with classical objects. This simulation can be used, e.g., to construct a computer with classical *hardware*, but with special *software*, in such a way that the whole machine simulates the behavior of a quantum object. One of the authors (A. A. G.) in his papers on QL[21−24], proposed calling this object a *macroscopic quantum computer*. In Chapter 8, we already discussed this idea in connection with QL.

9.2 VON NEUMANN'S THEOREM

Von Neumann[17] assumed that the states of a given quantum system are described by vectors in an appropriate Hilbert space and also the validity of the following axioms:

(i) There is a one to one correspondence between observables and Hermitian operators.

(ii) If to an observable R which take values r in a set $\mathcal{I} \subseteq \mathcal{R}$ there corresponds some Hermitian operator **R**, then to any function $f(r)$ there corresponds the operator f(R), given by

$$f(R) = \sum_{k=1}^{\infty} (\frac{1}{k!}) R^k \frac{d^k f}{dr^k}(0) \qquad (9.9)$$

(iii) If A and B are arbitrary observables and $a, b \in R$ then there is a linearity property for the expectation values in an ensemble of identical systems, i.e.,

$$\langle a\mathbf{A} + b\mathbf{B} \rangle = a\langle \mathbf{A} \rangle + b\langle \mathbf{B} \rangle \qquad (9.10)$$

which he used to prove several important results. The one we are interested in here is the the following:
There exists no dispersion free ensemble[9].
This result, of course, implies that there is no deterministic description of the ensemble.

We now give a *proof* of von Neumann's theorem for the simple case of a a spin 1/2 system (more precisely, a spin 1/2 particle) whose Hilbert space of interest, as we already know, is a complex two dimensional space. For a general discussion see[4,5,17].

Let us suppose that there exists some hidden parameter λ that defines the results of the observations of the spin property of our system. Let us make a series of measurements for identically prepared particles and let us call this series an *ensemble*.

Suppose that for the whole ensemble λ has the value λ_0 and that the ensemble has zero dispersion. Let us consider three observables R, S, T, to which correspond the 2×2 Hermitian matrices $\mathbf{R}, \mathbf{S}, \mathbf{T}$. Then, measurements of these observables in the ensemble with zero dispersion must lead to the same value λ_0 for λ. Then, R, S, T have fixed values $r(\lambda_0), s(\lambda_0), t(\lambda_0)$ which are respectively the eigenvalues of $\mathbf{R}, \mathbf{S}, \mathbf{T}$. Now, any 2×2 Hermitian matrix $\mathbf{X}$ can be written as,

$$\mathbf{X} = \alpha \mathbf{1}_2 + \vec{\beta} \cdot \vec{\sigma} \qquad (9.11)$$

where $\mathbf{1}_2$ is the unit 2×2 matrix, $\vec{\beta} = (\beta_1, \beta_2, \beta_3)$, $\beta_i \in \mathcal{R}$, $i = 1, 2, 3$ and $\vec{\sigma} \equiv (\sigma_1, \sigma_2, \sigma_3)$, where σ_i are the Pauli matrices. It follows that the eigenvalues of X are

$$\alpha \pm |\vec{\beta}|. \qquad (9.12)$$

Let $\mathbf{R} = \sigma_1$, $\mathbf{S} = \sigma_2$, $\mathbf{T} = \vec{\sigma} \cdot \vec{n}$, $\vec{n} = (1, 1, 0)$. Then $|\vec{n}| = \sqrt{2}$ and $\mathbf{T} = \mathbf{R} + \mathbf{S}$. Axiom (iii) implies that $\langle \mathbf{T} \rangle = \langle \mathbf{R} \rangle + \langle \mathbf{S} \rangle$, which leads to

$$t(\lambda_0) = r(\lambda_0) + s(\lambda_0). \qquad (9.13)$$

But it is impossible for Eq. (9.13) to be true, because the eigenvalues of $\mathbf{T}$ are $\pm\sqrt{2}$ and the eigenvalues of σ_1 and σ_2 are ± 1 and of course,

$$\pm\sqrt{2} \neq \pm 1 \pm 1. \qquad (9.14)$$

9.3 SPIN 1/2 HIDDEN VARIABLE'S MODEL

We now present a simple model of hidden variables[26] that reproduces all the results of quantum theory in the limited range for which it is applicable and which implies in a breakdown of von Neumann's axiom (iii), which expresses the linearity condition for expectation values.

The model is such that for a spin 1/2 system there corresponds the following *classical* model:

a) Particles with spin 1/2 are imitated by rotating spheres moving along the x-axis. Spin is the intrinsic angular momentum $\vec{\lambda}$ of each sphere and we suppose that it is oriented in some direction in the yz-plane.

b) To reproduce the probabilistic properties of the spin up $|z,1\rangle$ state, where

$$|z,1\rangle \equiv \begin{pmatrix} 1 \\ 0 \end{pmatrix} \qquad (9.15)$$

we suppose that we have a statistical ensemble of such spheres with a statistical distribution of spin vectors $\vec{\lambda}$ depending on the angle θ (between $\vec{\lambda}$ and the z-axis) given by:

$$\rho(\theta) = \begin{cases} \frac{1}{2}\cos\theta & \text{if } -\frac{\pi}{2} \leq \theta \leq \frac{\pi}{2} \\ 0 & \text{otherwise} \end{cases} \qquad (9.16)$$

c) The apparatus, measuring the observable X corresponding to the operator **X** given by Eq. (9.11) works according to the following rule: it measures the sign of the spin projection $\vec{\lambda}$ on the vector $\vec{\beta}$ and then multiplies the result with $|\vec{\beta}|$ and finally adds α. So, the result is always $\alpha \pm |\vec{\beta}|$ as in quantum mechanics. Due to Eq. (9.16), if σ_3 is measured ($\alpha = 0, \vec{\beta} = \vec{k}$, where $\vec{k}$ is the unit vector along the z-axis, also $\vec{i}$ and $\vec{j}$ are the unit vectors along the y and z axes respectively) then the angle between $\vec{\lambda}$ and $\vec{\beta}$ is always in the interval $(-\pi/2, \pi/2)$ and the projection on $\vec{\beta}$ is always nonnegative. But then, all measurements will give $\sigma_3 = +1$, which is the eigenvalue corresponding to the eigenvector $|z,1\rangle$ of σ_3.

d) It is also necessary to reproduce the quantum theory predictions concerning subsequent spin measurements in the same spin 1/2 system. To obtain this, it is enough to assume that each time that a spin measurement is made the $\vec{\lambda}$ vectors are redistributed in the statistical ensemble, with the probability function given by Eq. (9.16), but centered around the characteristic direction $\vec{\beta}$ of the new observable.

Let us now find the probabilities, p_1 and p_2 of getting the results $\alpha + |\vec{\beta}|, \alpha - |\vec{\beta}|$ if we measure $\mathbf{X} = \alpha \mathbf{1}_2 + \vec{\beta} \cdot \vec{\sigma}$. Observe that what is really measured is the sign of the scalar product $\vec{\lambda} \cdot \vec{\beta}$. Then p_1 and p_2 are equal to the actual a priori probabilities of the signs of $\vec{\lambda} \cdot \vec{\beta}$ which can be ± 1.

To find $p_1(\beta)$ we must calculate the integral of $\rho(\theta)$ over the dotted region of Fig.(9.1). Since the vectors $\vec{\lambda}$ can belong only to the upper semi-circle, we have

$$p_1(\beta) = \int_{-\frac{\pi}{2}+\theta_\beta}^{\frac{\pi}{2}} \frac{\cos\theta}{2} d\theta = \frac{1+\cos\theta_\beta}{2} \qquad (9.17)$$

Since $p(1) + p(2) = 1$, then $p_2(\beta) = \frac{1 - \cos\theta_\beta}{2}$.

The results obtained for $p_1(\beta)$ and $p_2(\beta)$ coincide with the results of quantum mechanics. The hidden variable is $\vec{\lambda}$. The value of any measurement $\alpha \mathbf{1}_2 + \vec{\alpha} \cdot \vec{\sigma}$ on the sphere is defined by the distribution probability of $\vec{\lambda}$ on the sphere.

It is easy to see that von Neumann's axiom (iii) is not valid here. Indeed, consider an ensemble with zero dispersion of particles with $\vec{\lambda}$ along the bisectors of the quadrant yz and measure for a part of the ensemble σ_2, for another part of the ensemble σ_3 and for yet another part of the ensemble $\vec{\sigma} \cdot \vec{n}$, with $\vec{n} = (0, 1, 1)$. Then $\sigma_2 + \sigma_3 = \vec{\sigma} \cdot \vec{n}$.

Now, due to the relation of $\vec{\lambda}$ with the directions $\vec{j}, \vec{k}$ and $\vec{m}$, $\langle \sigma_2 \rangle = +1$. Also, a measurement of σ_3 always gives $+1$, which gives $\langle \sigma_3 \rangle = +1$ and a measurement of $\vec{\sigma} \cdot \vec{n}$ always gives $+\sqrt{2}$, $\Longrightarrow \langle \vec{\sigma} \cdot \vec{n} \rangle = +\sqrt{2}$. But then, it is evident that $\langle \vec{\sigma} \cdot \vec{n} \rangle \neq \langle \sigma_2 \rangle + \langle \sigma_3 \rangle$, which means the breaking of von Neumann's axiom (iii).

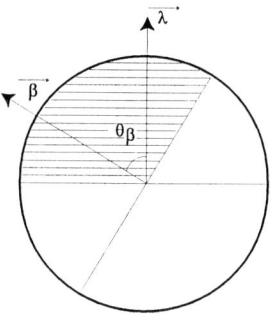

Figure 9.1. Integration region for calculation of $p_1(\beta)$.

The simple model presented above is enough to show that von Neumann's theorem, although mathematically exact, is irrelevant concerning the possibility of construction of hidden variables theories. It is very important to remark here that the realistic theory of quantum motion of spin 1/2 particles as developed by several authors, shows explicitly that modeling the measurement processes of the spin components using a Stern-Gerlach apparatus by means of an impulsive magnetic field implies that an unpolarized spin 1/2 beam is separated after passing through the apparatus in two non overlapping beams and just corresponds to what we usually call the spin up and down components. For details, the reader can consult the papers by Dewdney et al[27,28] to see how nicely the realistic theory of quantum motion fits the experimental data of Badurek's experiences[29] involving neutrons.[10]

9.4 GLEASON'S THEOREM. KOCHEN-SPECKER PARADOX

The construction of realistic *HVT* for a quantum system such that the Hilbert space has dimension $n \geq 3$ has some paradoxical features that do not occur in the case of the spin 1/2 system whose Hilbert space is two dimensional. It happens that in this case we find that the *predictions* concerning the results of measurements of a given observable A *depend* on the choice of another observable $B, C, \ldots$ that commutes with the first and that we measure simultaneously with A. Explicitly, what happens is that even supposing that a quantum system is *completely* described by the pair $(|\Psi\rangle, \lambda)$ the knowledge of the *value* of the hidden variable, in general does not permit predicting with certainty the result of a definite measurement of A, unless we specify the orthonormal basis used for writing the spectrum of A. Apparently, it is as if the value of the hidden variables λ change, simply because we decide to measure A simultaneously with another one of the observables $(B, C, ...)$ compatible with A. In this way, we come to the question of the uniqueness of the values of the hidden variables.

This phenomenon, as we have already said above, is called *contextual dependence*. In short, this is a result from Kochen-Specker[19] which follows from Gleason's theorem[18]. We now study the origin of the phenomenon and show that it is irrelevant concerning the possibility of construction of *HVT* of the first kind.

9.4.1 GLEASON'S THEOREM

Let $\{|\varphi_k\rangle\}$ be a countable orthonormal basis of $\mathcal{H}$ (dim $\mathcal{H} \geq 3$), the Hilbert space of a certain quantum system and let $\{\mathbf{P}_k = |\varphi_k\rangle\langle\varphi_k|\}$ be a basis for $\mathcal{L}^{\mathbf{P}}(\mathcal{H}) \subset \mathcal{H} \otimes \mathcal{H}^\dagger$, the lattice of projection operators. By definition, a *probability measure* on $\mathcal{L}^{\mathbf{P}}(\mathcal{H})$ is a mapping:

$$\mu : \mathcal{L}^{\mathbf{P}}(\mathcal{H}) \to [0,1] \in \mathcal{R} \tag{9.18}$$

which satisfies the standard axioms for a probability measure on each maximal Boolean sublattice of $\mathcal{L}^{\mathbf{P}}(\mathcal{H})$, the lattice of propositions of $\mathcal{H}$. i.e.,

$$\mu(\sum_k \mathbf{P}_k) = \sum_k \mu(\mathbf{P}_k) \tag{9.19}$$
$$\mu(\mathbf{0}) = 0, \mu(\mathbf{1}) = 1.$$

Suppose now that the $\mathbf{P}_i$ are projection operators corresponding to the spectrum of a given observable O, represented by the Hermitian operator $\mathbf{O}$. We impose that ,

$$\sum_k \mu(\mathbf{P}_k) = \mu(\sum_k (\mathbf{P}_k)) = \mu(\mathbf{1}) = 1. \tag{9.20}$$

Gleason proved that all probability measures on $\mathcal{H}$ such that dim$\mathcal{H} \geq 3$ are such that

$$\mu(\mathbf{P}_k) = Tr(\mathbf{P}_k \rho) \tag{9.21}$$

where ρ is the density matrix of the quantum system. The complete proof of Gleason's theorem is difficult and will not be presented here (see[18]). In order to prove the theorem, Gleason proved a lemma, which is the *key* to understand the *contextual dependence* phenomena mentioned above. To introduce Gleason's lemma, we need the concept of frame functions of weight w.

Let $\mathcal{H}' \subset \mathcal{H}$ be the subset of all normalized vectors of $\mathcal{H}$. A *frame function of weight w* is a mapping

$$f_w : \mathcal{H}' \ni |\psi\rangle \to f_w(|\psi\rangle) \in \mathcal{R}^+ \oplus \{0\} \tag{9.22}$$

such that, if $\{|\varphi_k\rangle\}$ is *any* orthonormal basis, then

$$\sum_k f_w(|\varphi_k\rangle) = w. \qquad (9.23)$$

The *frame functions* f_w *of weight* 1 will be denoted also by f, and in this case we have

$$0 \le f(|\varphi_k\rangle) \le 1. \qquad (9.24)$$

Let $\mathcal{H}'_3 \subseteq \mathcal{H}'$ be any three dimensional subspace of $\mathcal{H}'$, which is spanned by the vectors

$$|\Psi\rangle = \sum_{i=1}^{3} \alpha_i |\varphi_i\rangle \qquad (9.25)$$

and where $|\varphi_i\rangle \in \mathcal{H}', i = 1,2,3$ are any three orthonormal vectors. Let $\mathcal{R}'_3$ be any "real" subspace of orthonormal vectors of $\mathcal{H}'_3$, i.e., it is spanned by the orthonormal vectors $\{|e_i\rangle\}$ of $\mathcal{H}'_3$,

$$|v\rangle = \sum_{i=1}^{3} v_i |e_i\rangle, \quad \langle v|v\rangle = 1 \qquad (9.26)$$

where $v_i \in \mathcal{R}$.

Gleason's lemma: Any frame function

$$f: \mathcal{R}^3 \ni |v\rangle \mapsto f(|v\rangle) \in \mathcal{R}^+ + \{0\} \qquad (9.27)$$

must be a *continuous function* of the coefficients v_i defining $|v\rangle$ in Eq. (9.26).

Proof:

The proof is based on the fact that the set of normalized vectors in $\mathcal{R}_3$ is diffeomorphic to $\mathcal{S}^2$, the unit ball in $\mathcal{R}_3$, which is a connected manifold[11]. Then, each unit vector $|v\rangle$ can be parametrized by the usual spherical angles (ϑ, ϕ) and we denote by $F(\vartheta, \phi)$ the representative of $f(|v\rangle)$. Next, we consider two unit vectors, $|v'\rangle, |v''\rangle$ parametrized respectively (ϑ', ϕ') and (ϑ'', ϕ''). Now, since f is a frame function we have,

$$F(\vartheta, \phi) + F(\vartheta', \phi') + F(\vartheta'', \phi'') = 1 \qquad (9.28)$$

If we expand F in spherical harmonics,

$$F(\vartheta, \phi) = \sum_{lm} c_{lm} Y_{lm}(\vartheta, \phi) \qquad (9.29)$$

and use this expansion in Eq. (9.28) we discover immediately that $Y_{lm}(\vartheta, \phi)$ is a frame function for all pairs lm. We can then simply study the conditions for the functions

$$F_l(\vartheta, \phi) = \sum_{m=-l}^{l} c_m P_l^{|m|}(\cos \vartheta) e^{im\phi} \qquad (9.30)$$

HIDDEN VARIABLES THEORIES (HVT) 107

to be frame functions. We immediately realize that in Eq. (9.30) l can be only even, for otherwise it would be possible for F_l to change sign if the direction (ϑ, ϕ) is substituted for $(\pi - \vartheta, \phi + \pi)$ (its antipode) and obviously a frame function cannot do this. Then, in this case it is only necessary to consider the case for which $\vartheta = 0, \vartheta' = \pi/2 = \vartheta''$. In this way, we discover that $F_l(0, \phi) = c_0$, a constant. From this and the fact that F_l is a frame function, by using a well known result of harmonic analysis we find

$$F_l(\frac{\pi}{2}, \phi) + F_l(\frac{\pi}{2}, \phi + \frac{\pi}{2}) = \sum_{m=-l}^{l} c_m P_l^{|m|}(\cos\vartheta) e^{im\phi}(1 + i^m) = \text{constant} \quad (9.31)$$

which implies that the possible values of l are 0 and 2, because, as well known[32], the representations of the rotation group by spherical harmonics are *irreducible*. This means that if $l \geq 4$, we can take $c_4 = 0$ due to the arbitrariness in the choice of the polar axis. This implies that all the $c_m = 0$.

With a simple coordinate transformation, we can write the $F(\vartheta, \phi) \equiv f(|v\rangle)$ in terms of the Cartesian components of $|v\rangle$ and we arrive at the conclusion that the *possible form* of the frame functions are:

$$f(|v\rangle) = \sum_{i,j=1}^{3} t_{ij} v_i v_j \quad (9.32)$$

where the set (t_{ij}) defines a nonnegative matrix with unit trace.

The proof of Gleason's lemma for Hilbert spaces of dimension greater then 3 is more complicated, but the following result is true: all frame functions of weight one like $f_{w=1} \equiv f$ in eq(9.22) are of the form,

$$f(|\Psi\rangle) = f(\sum_{i=1}^{3} \alpha_i |\varphi_i\rangle) = \sum_{i,j=1}^{3} m_{ij} \alpha_i^* \alpha_j \quad (9.33)$$

where the set (m_{ij}) defines a nonnegative matrix with unit trace.

Nonexistence of truth values functions. Suppose that there exists a mapping f called truth value function which gives to each basis of orthonormal vectors the value 1 to one of the vectors of the basis and 0 to the others. Let F be the representative in S^2 of our frame function f. Then,

$$F : S^2 \to \{0, 1\} \quad (9.34)$$

such that for each $p \in S^2$, $F(p)$ has value 1 or 0.

Now, by Gleason's lemma, the map must be continuous. Then the inverse images of the open sets $\{0\}$ and $\{1\}$ must correspond respectively to the open sets $F^{-1}(\{0\})$ and $F^{-1}(\{1\})$ on S^2. Since any $p \in S^2$ is mapped in 1 or 0, it follows that

$$F^{-1}(\{0\}) = \overline{F^{-1}(\{1\})}, \quad F^{-1}(\{0\}) \cap F^{-1}(\{1\}) = \emptyset \quad (9.35)$$

where $\overline{F^{-1}(\{1\})}$ is the set theoretical complement. Then, by the definition of complement we must have

$$\begin{aligned} S^2 &= F^{-1}(\{1\}) \cup \overline{F^{-1}(\{1\})} \\ &= F^{-1}(\{0\}) \cup F^{-1}(\{1\}) \end{aligned} \quad (9.36)$$

which is a contradiction, since S^2 is connected. It follows that no frame function like f in Eq. (9.27) can be a truth value function.

We can generalize trivially the above result for the frame functions f defined by eq(9.22), since $\mathcal{H}'$ is diffeomorphic to the hyperball S^{2n-1}, which is a connected manifold. Then, it follows that there are no truth functions of the kind

$$f : \mathcal{H}' \to \{0,1\}, \quad \sum_k f(|\varphi_k\rangle) = 1. \tag{9.37}$$

9.4.2 KOCHEN-SPECKER THEOREM

The Kochen-Specker theorem is the statement that for a Hilbert space $\mathcal{H}$, such that $\dim \mathcal{H} \geq 3$, there are no truth value functions on the lattice $\mathcal{L}^\mathbf{P}(\mathcal{H}) \subset \mathcal{H} \otimes \mathcal{H}^\dagger$, i.e., there are no functions

$$V : \mathcal{H} \otimes \mathcal{H}^\dagger \supset \mathcal{L}^\mathbf{P}(\mathcal{H}) \to \{0,1\}, \quad \sum_k V(\mathbf{P}_k) = 1 \tag{9.38}$$

such that $V(\mathbf{P}_i) = 1$ for some $\mathbf{P}_i$ in the set $\{\mathbf{P}_k = \sum_k |\varphi_k\rangle\langle\varphi_k|\}$ and $V(\mathbf{P}_j) = 0$, if $j \neq i$.

This theorem follows at once from the generalization of Gleason's lemma given in Eq. (9.33), and the nonexistence of truth value functions once we take into account that the frame functions f satisfy

$$f(e^{i\beta}|\Psi\rangle) = f(|\Psi\rangle), \quad \beta \in \mathcal{R} \tag{9.39}$$

and that the lattice $\mathcal{L}_P^\mathcal{H}$ (which as a vector space over $\mathcal{C}$ has the same dimension of $\mathcal{H}'$) is diffeomorphic to the quotient space S^{2n-1}/S^1. Then the representative function F of f in S^{2n-1} passes to the quotient defining a function

$$F_\mu : S^{2n-1}/S^1 \to \in \mathcal{R}^+ + \{0\} \tag{9.40}$$

which is the representative of the function

$$\mu : \mathcal{L}^\mathbf{P}(\mathcal{H}) \to \in \mathcal{R}^+ + \{0\} \tag{9.41}$$

such that

$$\mu(\mathbf{P}_{|\Psi\rangle}) \equiv f(|\Psi\rangle), \quad \mathbf{P}_{|\Psi\rangle} = |\Psi\rangle\langle\Psi|. \tag{9.42}$$

It follows that no truth value function V exists on $\mathcal{L}^\mathbf{P}(\mathcal{H})$.

9.4.3 CONTEXTUALITY IN HVT AND THE MERMIN EXAMPLE

We recall that a Hermitian operator $\mathbf{O}$ that represents a physical observable O is said to be maximal if it is nondegenerate. In this case, there is only one basis that diagonalizes it. Physically, this means that this basis corresponds to a maximal measurement of the physical observable O. When the operator $\mathbf{O}$ is degenerate, there are many distinct bases that can diagonalize it. These bases define inequivalent measurement processes and because *all* are called measurements of $\mathbf{O}$, this gives a non trivial ambiguity to the word measurement.

A set of commuting (also said compatible) operators $\{\mathbf{O}_m\}$, $m = 1, 2, \ldots M$ is said to be *complete*, if there is a basis that diagonalizes all of them. A complete set is said to be *maximal* if given another operator $\mathbf{A}$ that commutes with all the $\mathbf{O}_m$, the new set that includes A is not complete. Then, we see that the simultaneous measurement of observables corresponding to a maximal complete set of operators is equivalent to the measurement of an observable represented by a maximal operator.

A fundamental assumption of quantum theory which is called *independence of context* is the following: if a quantum system is in some state $|\Psi\rangle$ and for which $\{\mathbf{O}_m\}, m = 1, 2 \ldots M$ is a maximal complete set, then the result of the measurement of, let us say, $\mathbf{O}_1$, does not depend on the fact that we choose to measure $\mathbf{O}_1$ simultaneously with a given $\mathbf{O}_k$ ($k = 2, 3, \ldots M$) or to measure $\mathbf{O}_1$, together with a given $\mathbf{O}_j, j \neq k$ ($j = 2, 3, \ldots M$).

Suppose now, that we have a situation where there is some operator $\mathbf{A}$ that can be written as a function g, let as say, of two different operators $\mathbf{O}_1, \mathbf{O}_2$, of the maximal complete set $\{\mathbf{O}_m\}$, or as a function h of the operators $\mathbf{O}_3, \mathbf{O}_4$. Then, if our system is in some state $|\Psi\rangle$, such that

$$\mathbf{O}_1|\Psi\rangle = a|\Psi\rangle, \quad \mathbf{O}_2|\Psi\rangle = b|\Psi\rangle, \quad (9.43)$$
$$\mathbf{O}_3|\Psi\rangle = c|\Psi\rangle, \quad \mathbf{O}_4|\Psi\rangle = d|\Psi\rangle,$$

then

$$\mathbf{A}|\Psi\rangle = g(\mathbf{O}_1, \mathbf{O}_2)|\Psi\rangle = g(a, b)|\Psi\rangle \quad (9.44)$$
$$\mathbf{A}|\Psi\rangle = h(\mathbf{O}_3, \mathbf{O}_4)|\Psi\rangle = h(c, d)|\Psi\rangle$$

At first sight, it seems reasonable to suppose that also in the case that the quantum system is in a state $|\Phi\rangle$ which is not an eigenvector of $\mathbf{O}_i, i = 1, 2, 3, 4$ and even if these operators are not measured the possible results of their measurements (if made) satisfy the same functional relation as the one satisfied by the operators. More explicitly, introduce the functions

$$V(\mathbf{A})^{|\Phi\rangle}, V(\mathbf{O}_i)^{|\Phi\rangle}, \quad i = 1, 2, 3, 4 \quad (9.45)$$

that give the values of the observables $A, \mathbf{O}_i$ in the state $|\Phi\rangle$. The above assumption, known in literature as the *functional composition principle (FUNC)*, then can be written

$$V(\mathbf{A})^{|\Phi\rangle} = g(V(\mathbf{O}_1)^{|\Phi\rangle}, V(\mathbf{O}_2)^{|\Phi\rangle}) = h(V(\mathbf{O}_3)^{|\Phi\rangle}, V(\mathbf{O}_4)^{|\Phi\rangle}). \quad (9.46)$$

The *FUNC rule* has been introduced by Jauch and Piron[33], and by Kochen and Specker[19] as a *natural postulate* expressing the realism behind the *HVT*. Now, this postulate is incompatible with the independence of context assumed by quantum theory. Indeed, supposing that both postulates are true, we get contradictions. These contradictions arise due to the Kochen-Specker theorem studied above, a consequence of Gleason's theorem. We now illustrate the situation presenting Mermin's example[34].

Consider a quantum system consisting of two spin 1/2 particles. The Hilbert space of the system is $\mathcal{H}^1 \otimes \mathcal{H}^2$, the spin operators corresponding to each particle are $\sigma_i^{(1)} \otimes \mathbf{1}_2$ and $\mathbf{1}_1 \otimes \sigma_i^{(2)}$, as we already know from Chapter 2. Consider equation (9.47) where all operators in the third row (column) are the product of the operators of the *commuting operators* in the the other two rows (columns), except for the third row where a positive sign is needed instead of a negative sign.

$$\begin{array}{cccc}
\mathbf{1}_1 \otimes \sigma_3^{(2)} & \sigma_2^{(1)} \otimes \mathbf{1}_2 & \sigma_2^{(1)} \otimes \sigma_3^{(2)} & 1 \\
\sigma_1^{(1)} \otimes \mathbf{1}_2 & \mathbf{1}_1 \otimes \sigma_1^{(2)} & \sigma_1^{(1)} \otimes \sigma_1^{(2)} & 1 \\
\sigma_1^{(1)} \otimes \sigma_3^{(2)} & \sigma_3^{(1)} \otimes \sigma_1^{(2)} & -\sigma_2^{(1)} \otimes \sigma_2^{(2)} & -1 \\
1 & 1 & 1 &
\end{array} \quad (9.47)$$

If we admit the validity of *FUNC* as applying to all the nine observables in the table, then we get a contradiction. Indeed, using Eq. (9.46), we can write six equations for the product of three observables in each row (column). The result is shown in the last row (column) of Eq. (9.47). Thus, we have

$$\begin{aligned}
V(\mathbf{1}_1 \otimes \sigma_3^{(2)})^{|\Phi\rangle} V(\sigma_2^{(1)} \otimes \mathbf{1}_2)^{|\Phi\rangle} V(\sigma_2^{(1)} \otimes \sigma_3^{(2)})^{|\Phi\rangle} &= 1 \\
V(\sigma_1^{(1)} \otimes \mathbf{1}_2)^{|\Phi\rangle} V(\mathbf{1}_1 \otimes \sigma_1^{(2)})^{|\Phi\rangle} V(\sigma_1^{(1)} \otimes \sigma_1^{(2)})^{|\Phi\rangle} &= 1 \\
V(\sigma_1^{(1)} \otimes \sigma_3^{(2)})^{|\Phi\rangle} V(\sigma_3^{(1)} \otimes \sigma_1^{(2)})^{|\Phi\rangle} V(-\sigma_2^{(1)} \otimes \sigma_2^{(2)})^{|\Phi\rangle} &= -1 \\
V(\mathbf{1}_1 \otimes \sigma_3^{(2)})^{|\Phi\rangle} V(\sigma_1^{(1)} \otimes \mathbf{1}_2)^{|\Phi\rangle} V(\sigma_1^{(1)} \otimes \sigma_3^{(2)})^{|\Phi\rangle} &= 1 \\
V(\sigma_2^{(1)} \otimes \mathbf{1}_2)^{|\Phi\rangle} V(\mathbf{1}_1 \otimes \sigma_1^{(2)})^{|\Phi\rangle} V(\sigma_3^{(1)} \otimes \sigma_1^{(2)})^{|\Phi\rangle} &= 1 \\
V(\sigma_2^{(1)} \otimes \sigma_3^{(2)})^{|\Phi\rangle} V(\sigma_1^{(1)} \otimes \sigma_1^{(2)})^{|\Phi\rangle} V(-\sigma_2^{(1)} \otimes \sigma_2^{(2)})^{|\Phi\rangle} &= 1 \quad (9.48)
\end{aligned}$$

Multiplying all these equations, we see that the left member is 1 since all values appear as a square, but the right member is -1.

9.4.4 KOCHEN-SPECKER PARADOX

Consider a spin 1 quantum system[12]. Let $\mathbf{J}_i$, $i = 1, 2, 3$ be the angular momentum operators and let

$$\mathbf{J}^2 = \sum_i \mathbf{J}_i^2 \quad (9.49)$$

Consider the vectors $\varphi_m \equiv |j, m\rangle$, (where $j = 1$ and m is an integer) such that

$$\begin{aligned}
\mathbf{J}^2 \varphi_m &= j(j+1)\varphi_m, \; j = 1 \\
\mathbf{J}_3 \varphi_m &= \begin{cases} m\varphi_m & \text{if } |m| < j \\ 0 & \text{if } |m| > j \end{cases}
\end{aligned} \quad (9.50)$$

$$(\mathbf{J}_1^2 \mathbf{J}_2^2 - \mathbf{J}_2^2 \mathbf{J}_1^2)\varphi_m =$$
$$(1-m)[(j+m)(j+m-1)(j-m+2)(j-m+1)]^{\frac{1}{2}}\varphi_{m-2}$$
$$+ (1+m)[(j+m+2)(j+m+1)(j-m)(j-m-1)]^{\frac{1}{2}}\varphi_{m+2} \quad (9.51)$$

Then, for $j = 1$, $\varphi_{m\pm 2}$ is different from zero only if $m = \pm 1$, but then $(1 \pm m)$ is equal to zero and it follows that $\mathbf{J}_1^2$ commutes with $\mathbf{J}_2^2$ and by symmetry it commutes with $\mathbf{J}_3^2$ and then we see that it commutes with $\mathbf{J}^2$. It follows that $\mathbf{J}^2$ and $\mathbf{J}_i^2, i = 1, 2, 3$ have simultaneous eigenvectors. Now, $\mathbf{J}^2$ has eigenvalue 2 and $\mathbf{J}_i^2$ can have eigenvalues 1 or 0. It follows that two of them must have eigenvalues equal to 1 and the other equal to zero.

Kochen and Specker proposed an experiment with orthohelium in an electric field with rhombic symmetry to verify the above prediction. The idea is based on the fact that the Hamiltonian for the perturbation energy of the atom is given by

$$\mathbf{H}_S = a\mathbf{J}_1^2 + b\mathbf{J}_2^2 + c\mathbf{J}_3^2 \qquad (9.52)$$

where in this case $\mathbf{J}_i = \mathbf{S}_i$. It is clear that the perturbation energy which depends on each of the three eigenvalues J_i^2 of the operators $\mathbf{J}_i^2$ is zero. Now, if the orthohelium is excited in the state 2^3P_0, then in absence of the rhombic field, it will decay to any 2^3S_1 state by emission of a photon. By applying the rhombic electric field, we can measure where the atom ends with its spin in one of the planes 23 or 31 or 12. To proceed, observe that for spin 1 systems, the operators

$$\mathbf{P}_i = 1 - \mathbf{J}_i^2 \qquad (9.53)$$

(as can be easily verified) are projection operators. Each one corresponds to the question: is the value of $\mathbf{J}_i^2 = 0$?

Now, the main objective of *HVT* for the above problem is *to predict* for our quantum system *completely* described by the pair ($|\Psi\rangle, \lambda$) and for each arbitrarily given orthonormal frame $\{\vec{e}_i\}$ in Euclidean space, which is the direction $\vec{e}_i$ for which the value of $\mathbf{P}_i$ is 1.

Now, Kochen and Specker showed that we obtain a paradox for the problem of the orthohelium if we supposed that HVT implies that from the knowledge of $(|\Psi\rangle, \lambda,)$ there exists a function

$$V(|\Psi\rangle, \lambda, \mathbf{P}_m) \equiv V(\mathbf{P}_m) = V(1 - (\vec{m} \cdot \mathbf{J})^2), \ \mathbf{P}_m = 1 - \vec{m} \cdot \mathbf{J}, \ \vec{m} \cdot \vec{m} = 1, \qquad (9.54)$$

with values 1 or 0 corresponding to the cases where $\mathbf{J}_m^2$ is 0 or 1 in a measurement, *without* specifying the triad to which $\vec{m}$ belongs. This is so, because we can identify $\vec{m}$ with one $\vec{e}_i$, let us say, $\vec{m} = \vec{e}_1$ and to satisfy all the requirements of our problem we certainly must have

$$\sum_i V(\mathbf{P}_i) = 1. \qquad (9.55)$$

A function like this would be a truth value function. However, now, we are under the conditions of the Gleason and the Kochen-Specker theorems that prohibit the existence of these functions.

We arrive at the conclusion that any consistent *HVT* can predict the value of $\mathbf{P}_m$ only if we fix an orthonormal basis in the physical Euclidean space that permits univocally fixing a basis in the space $\mathcal{L}_P^\mathcal{H}$. We then have shown that any consistent *HVT* must be context-dependent, this being the reason why we have defined that *HVT* of the first kind obeys this condition. Then, to suppose the existence of a *HVT* that is context independent (theories of the zeroth kind), is to suppose the existence of *no HVT*. A review of other non collapses interpretations of quantum theory different from the *BBT* can be found in the recent book by Bub[35].

We end with the observation that Bell arrived at the Kochen-Specker result in an independent way. A discussion of Bell's locality theorem and his version of the Kochen-Specker theorem can be found in [36].

Notes

1. However, if you have felt uncomfortable with our subject, you are in good company, as that of Einstein, de Broglie and Schrödinger, only to mention some of the founding fathers. A good history of the discomfort with the orthodox interpretation of quantum theory can be found in Jammer's book[1]. De Broglie's books[2,3] are also very instructive and also Selleri's book[4]. On the other hand, Bohr, Heisenberg and von Neumann are not bad company for those who do not fell uncomfortable.
2. A statistical ensemble is said to be homogeneous when all its subensembles give the same expectation value for all observables. We always suppose that our ensembles are homogeneous, even when this fact is not explicitly stated.
3. The mathematical nature of λ depends on the specific hidden variable theory.
4. In this sense *HVT* are said to be realistic interpretations of quantum theory.
5. We recall that philosophically speaking, ontology is especially interested in the nature of the being, i.e., of reality in abstraction, and is only secondarily interested in the possible ways to obtain the knowledge about the being.
6. De Broglie- Bohm theory is now presented with details in two recent books, one by Bohm and Hilley[6] and the other by Holland[7].
7. This is a very common situation that occurs in quantum theory, e.g., when **O** has degenerate eigenvalues. We discuss this issue below in details.
8. Well, not for everybody, since there are people[25] who still have doubts regarding Aspect's experiments and since it may eventually be the case that they are right.
9. We recall that the dispersion of a given observable A, represented by the operator **A**, is defined by $\sigma_A \equiv \triangle A = \langle \mathbf{A} \rangle^2 - \langle \mathbf{A}^2 \rangle$. An ensemble is said to be dispersion free if $\sigma_A = 0$ for all observables of the characteristic system of the ensemble.
10. A Clifford algebra approach to quantum theory reveals even a profound connection between the classical and quantum spin1/2 motions. In this connection see [30,31].
11. A connected manifold is a manifold that cannot be represented as the union of disjoint open sets.
12. Here we use units such that $\hbar = 1$

References

[1] M. Jammer, *The Philosophy of Quantum Mechanics* (J. Wiley & Sons, New York, 1974).

[2] L. de Broglie, *Non-Linear Wave Mechanics. A Causal Interpretation* (Elsevier Publishing Co., Amsterdam, 1960).

[3] L. de Broglie, *Heisenberg's Uncertainties Principles and the Probabilistic Interpretation of Wave Mechanics* (Kluwer Academic Publishers, Dordrecht, 1990).

[4] F. Selleri, *Quantum Paradoxes and Physical Reality* (Kluwer Academic Publishers, Dordrecht, 1990).

[5] F. J. Belinfante, *A Survey of Hidden-Variables Theories* (Pergamon Press, Oxford, 1973).

[6] D. Bohm and B. J. Hilley, *Undivided Universe* (Routledge, London and New York,1993).

[7] P. R. Holland, *The Quantum Theory of Motion* (Cambridge University Press, 1993).

[8] Ph. Gueret and J. P. Vigier, De Broglie's wave particle duality in the stochastic interpretation of quantum mechanics: a testable physical assumption, *Found. Phys.* **12**(11), 1057–1083 (1982).

[9] C. Papaliolios, Experimental test of a hidden-variable quantum theory, *Phys. Rev. Lett.* **18**, 622–625 (1967).

[10] L. de Broglie, Sur la possibilité de relier les phénomènes d'interférence et de diffraction à la théorie des quanta de lumière, *Compt. Rend.* **183**, 447–448 (1926).

[11] L. de Broglie, La structure de la matière et du rayonnement et la mécanique ondulatoire, *Compt. Rend.* **184**, 273–274 (1927).

[12] L. de Broglie, La mécanique ondulatoire et la structure atomique de la matiére et du rayonnement, *J. Phys. Radium* **8**(5), 225–241 (1927).

[13] A. Aspect, J. Dallibard and G. Roger, Experimental tests of realistic local theories via Bell's theorem, *Phys. Rev. Lett.* **47**(7), 460–463 (1981).

[14] A. Aspect, P. Grangier and D. Roger, Experimental realization of Einstein-Podolski-Rosen gedankenexperiment—a new violation of Bell inequalities, **49**(2), 91–94 (1982).

[15] A. Aspect, J. Dallibard and G. Roger, Experimental test of Bell inequalities using time varying analyzers, *Phys. Rev. Lett* **49**(25), 1804–1807 (1982).

[16] A. Aspect and P. Grangier, About resonant scattering and other hypothetical effects in the Orsay atomic cascade experiment tests of Bell inequalities—a discussion and some new experimental data, *Nuovo Cimento* **43**(8), 345–348 (1985).

[17] J. von Neumann, *Mathematical Foundations of Quantum Mechanics* (Princeton University Press, Princeton, NJ,1955).

[18] A. M. Gleason, Measures on the closed subspaces of a Hilbert space, *J. of Rat. Mech. and Analysis* **6**, 885–893 (1957).

[19] S. Kochen and E. P. Specker, The problem of hidden variables in quantum mechanics, *J. of Rat. Mech. and Analysis* **17**, 59–87 (1967).

[20] J. M. Jauch and C. Piron, Can hidden variables be excluded in quantum mechanics, *Helv. Phys. Acta* **36**(7), 827–837 (1963).

[21] A. A. Grib and R. R. Zapatrin, Automata simulating quantum logics, *Int. J. Theor. Phys.* **29**(2), 113–123 (1990).

[22] A. A. Grib and R. R. Zapatrin, Quantum logical description of two-particle systems, *Int. J. Theor. Phys.* **30**(7), 949–959 (1991).

[23] A. A. Grib, Quantum logic interpretation of quantum mechanics—the role of time, *Int. J. Theor. Phys.* **32**(12), 2389–2400 (1993)

[24] A. A. Grib and R. R. Zapatrin, Topologimeter and the problem of physical interpretation of the topology lattice, *Int. J. Theor. Phys.* **35**(3), 593–604 (1996)

[25] J. P. Wesley, *Classical Quantum Theory* (Benjamin Wesley Publisher, Blumberg, Germany, 1996).

[26] D. Bohm, R. Schiller and J. Tiomno, A causal interpretation of Pauli equation, *Nuovo Cimento* **1**(supp.), 48–66 (1955).

[27] C. Dewdney, P. R. Holland and A. Kypryanidis, What happens in a spin measurement, *Phys. Lett. A* **119**(6), 259267 (1986).

[28] C. Dewdney, P. R. Holland and A. Krypianidis, A causal account of nonlocal Einstein-Podolski-Rosen spin correlations, *J. Phys. A—Math. and Gen.* **20**(14), 4732–4732 (1987).

[29] G. Badurek, H. Rauch and D. Tuppinger, Neutron interferometric double-resonance experiment *Phys. Rev. A* **34**(4), 2600–2608 (1986).

[30] W. E. Baylis (ed.), *Clifford (Geometric) Algebras with Applications in Physics, Mathematics and Engineering*, Birkhäuser, Boston, MA, 1996.

[31] W. A. Rodrigues, Jr., J. Vaz, Jr. and M. Pavsic The Clifford bundle and the dynamics of the superparticle, *Banach Center Publ., Polish Acad. Sci.* **37**, 295–314 (1996).

[32] W. Miller, Jr., *Symmetry Groups and their Applications* (Academic Press, New York, 1972).

[33] J. M. Jauch and C. Piron, On the structure of quantum proposition system, *Helv. Phys. Acta* **42**, 842–848 (1969).

[34] N. D. Mermin, Simple unified form for the major no-hidden-variables theorems, *Phys. Rev. Lett.* **65**(27), 3373–3376 (1990).

[35] J. Bub, *Interpreting the Quantum World* (Cambridge University Press, 1997).

[36] H. R. Brown, in: *Bell's Theorem and the Foundations of Modern Physics*, edited by A. van der Merwe, F. Selleri and G. Tarrozi (World Scientific Publ. Co., Singapore, 1992), pp. 104—116.

10
DE BROGLIE-BOHM NON RELATIVISTIC HVT

The recent developments of the de Broglie-Bohm *HVT (BBT)*, to which Bohm and Hilley[1] refer as the "Ontological Interpretation of Quantum Theory" and Holland[2] calls "The Quantum Theory of Motion," will be discussed in what follows. The main thesis of the *BBT* is that *reality* is composed of *beables*, these being the *fermion particles* and their *quantum waves* and the *boson fields* and their *quantum waves*. In this Chapter, we discuss the main aspects of the non relativistic *one particle* and *many particles* theories and in the next Chapter we study the relativistic *one particle* and *many particles* fermion theories and the boson quantum field theories[1].

10.1 NON RELATIVISTIC DE BROGLIE-BOHM THEORY

The main idea of the non relativistic *BBT* is that the trajectories of microparticles, contrary to what the standard quantum theory says, have *real* existence. The motion of particles along these trajectories is described by classical theory, but the wave function *guides* this motion through a special *quantum potential*, depending on the wave function. The pilot wave conjecture has been proposed by de Broglie[11]. The wave guides the motion of the particle, *instructing* it how to move. In this way, the role of the wave function can be compared with the *program* or *software* in a computer which commands the hardware. Formally, we proceed as follows. We suppose that associated to our one particle system (with mass m) moving in a potential $V(t,\vec{x})$, there is a wave function $\Psi(t,\vec{x})$ (where t is the time parameter and $\vec{x}$ denotes the coordinates of the Euclidean space where the particle moves) which satisfies the usual

Schrödinger equation,

$$i\hbar \frac{\partial \Psi}{\partial t} = -\frac{\hbar^2}{2m}\nabla^2 \Psi + V\Psi \quad (10.1)$$

We write,

$$\Psi = R\exp(iS/\hbar). \quad (10.2)$$

Substituting Eq. (10.2) into Eq. (10.1), we get the following equations for functions R and S,

$$\frac{\partial S}{\partial t} + \frac{(\nabla S)^2}{2m} + V + Q = 0 \quad (10.3)$$

$$\frac{\partial \rho}{\partial t} + \nabla \cdot \left(\frac{\rho \nabla S}{m}\right) = 0 \quad (10.4)$$

where,

$$Q = -\frac{\hbar^2}{2m}\frac{\nabla^2 R}{R} \quad (10.5)$$

and

$$R = |\Psi|, \quad \rho = |\Psi|^2 = R^2. \quad (10.6)$$

If S in Eq. (10.3) is interpreted as the action, then this equation has the form of the classical Hamilton-Jacobi equation for a particle moving in a potential, $V + Q$. Q is called the *quantum potential*. Equation (10.4) describes the motion of a liquid with density ρ and velocity field $\nabla S/m$.

The quantum potential Q depends on the form of the wave function and does not change if Ψ is multiplied by an arbitrary complex constant. This feature is considered important, and we can say that the quantum potential describes *information* contained in the wave function, directly guiding the motion of the particle (or of the energy). This guidance is compared by Bohm and Hilley to the motion of a car under the influence of a green or red light. The physical action of the signal is very small, but the information contained in it, leads to the change of the motion of the car.

The hidden variable, that we denoted by λ in the last Chapter, is the real position of the particle in Euclidean space. The particle follows one of the curves

$$\sigma : R \to R^3, t \longmapsto \sigma(t) \quad (10.7)$$

determined by one of the integral lines of the velocity field $\nabla S/m$. We denote the coordinate functions of each one of the integral lines by $\vec{x}(\sigma(t)) \equiv \vec{x}(t)$. Then the velocity of the particle following the trajectory σ is given by the guidance formula,

$$\vec{v} = \frac{d}{dt}\vec{x}(\sigma(t)) = \frac{1}{m}\vec{P}|_\sigma = \vec{V}|_\sigma \quad (10.8)$$

$$\vec{V} = \frac{1}{m}\vec{P} = \frac{\nabla S}{m} = \frac{\vec{j}_\Psi}{\rho} \quad (10.9)$$

where,

$$\vec{j}_\Psi = -\frac{i\hbar}{2m}\left(\Psi^*\nabla\Psi - \Psi\nabla\Psi^*\right) \quad (10.10)$$

is the Schrödinger current. It is supposed that the wave function $\Psi(t,\vec{x})$ is a partial description of an ensemble of particles which is here fully described by the pair $[\vec{x}(\sigma(t)), \Psi(t,\vec{x})]$, where each trajectory σ occurs with probability density ρ. We recall that in order to solve Eq. (10.8) it is necessary to specify the initial position $\vec{x}_0$. The wave function Ψ is a single valued function. From this it follows that the S function is clearly multivalued since S and $S + n\hbar$ give rise to the same Ψ. However, $\vec{P}$ is a single valued function, which implies that $\vec{x}(t)$ is also single valued.

10.1.1 STATIONARY STATES

In a stationary state the wave function can be written as,

$$\Psi(t,\vec{x}) = \psi_0(\vec{x}) \exp\left(-iEt/\hbar\right). \quad (10.11)$$

The quantum Hamilton-Jacobi equation gives,

$$E = \frac{(\nabla S)^2}{2m} + V + Q. \quad (10.12)$$

If we apply this equation, let us say, to the s-state of a hydrogen atom, since in this case ψ_0 is real, it follows that $S = 0$ and then

$$\vec{P} = 0, \; E = V + Q. \quad (10.13)$$

This equation tells us that if the BBT is correct, then in the s-state, the electron is at rest! This is not what we expect from classical physics, but at least it seems to indicate that in a state of dynamic equilibrium, the quantum potential cancels out the space variation of the potential V and as a consequence the particle does not radiate nor fall into the nucleus. At first sight, this explanation presented in [1,2] seems satisfactory. However, if we take into account that Schrödinger's equation must be obtained as the limit of Pauli's equation (which is the non-relativistic limit of the Dirac equation) when spin is frozen we immediately realize[13] that we can not use Schrödinger's current (Eq. (10.10)) for this problem, because it is not the appropriate non-relativistic limit of the Dirac current when the spin is *frozen*. When the correct limit is taken into account there is an additional term in the correct current which implies that the electron is not at rest. This remark shows, in particular, how dangerous it is to discuss the meaning of different physical theories without their relationships taking appropriately into account (for more details, see[13]).

10.1.2 NON CONSERVATION OF ENERGY AND MOMENTUM IN NON STATIONARY STATES

In a non stationary state, different from the stationary case, the energy of the particle, $\partial S/\partial t$ will not be conserved in detail, as can be easily verified. However it is conserved

118   NONLOCALITY IN QUANTUM PHYSICS

in the average. Indeed, we have,

$$\langle E \rangle = \int \rho \frac{\partial S}{\partial t} d^3 x$$
$$= -\int R^2 \left[\frac{(\nabla S)^2}{2m} + V - \frac{\hbar^2}{2m} \frac{\nabla^2 R}{R} \right] d^3 x \quad (10.14)$$

Using Eq. (10.2), we can write Eq. (10.14) as

$$\langle E \rangle = \int d^3 x \left[\frac{-\hbar^2}{2m} \Psi^* \nabla^2 \Psi + \Psi^* V \Psi \right] \quad (10.15)$$

which we recognize as the mean value of the Hamiltonian operator in the standard interpretation of quantum theory. This result, of course, implies in energy conservation in the classical limit where it is possible to neglect Heisenberg's uncertainty principle. However, we cannot avoid the question: from where does the fluctuation energy (and also momentum) come? There is no coherent answer to this question in the theory, because according to Bohm and Hilley, the wave function Ψ is an information field and does not carry energy[2]. They then suggests that the energy and momentum fluctuations originate in the vacuum fluctuations, but they do not present any convincing mathematical argument that this may indeed be the case...

The applications of the *BBT* for the interference pattern in a double slit experiment for the tunneling problem and the Aharanov-Bohm effect are worth mentioning, because they give us (contrary to the standard interpretation of quantum theory) a very clear picture of these phenomena (see[1]).

10.2   THE MANY PARTICLE SYSTEM

In *BBT* an n-particle system is defined by:

(i) A wave function $\Psi(t, \vec{x}_1, ..., \vec{x}_n)$ (where $\vec{x}_i, i = 1, ..., n$ are Cartesian coordinates of the $3n$ dimensional configuration space $\mathcal{C}_{3n}$ of the system) which obeys the n-particle Schrödinger equation,

$$i\hbar \frac{\partial \Psi}{\partial t} = \left[\sum_{i=1}^{n} -\frac{\hbar^2}{2m_i} \nabla_i + V \right] \Psi \quad (10.16)$$

where ∇_i acts on the variable $\vec{x}_i$, and V is in general a function of $(t, \vec{x}_1, ..., \vec{x}_n)$.

(ii) A set of point particles following trajectories $\sigma_i : \mathcal{R} \to \mathcal{R}^3, i = 1, ..., n$. Note that a trajectory in $\mathcal{C}_{3n}$ is equivalent to n particle trajectories in $\mathcal{R}^3$.

Writing,

$$\Psi = R \exp(iS/\hbar) \quad (10.17)$$

we get

$$\frac{\partial S}{\partial t} + \sum_{i=1}^{n} \frac{(\nabla_i S)^2}{2m_i} + V + Q = 0 \quad (10.18)$$

$$\frac{\partial R^2}{\partial t} + \sum_{i=1}^{n} -\hbar^2 \nabla_i^2 R / m_i R = 0 \quad (10.19)$$

where the quantum potential Q, which in general is a function of $(t, \vec{x}_1, ..., \vec{x}_n)$ is given by

$$Q = \nabla_i \cdot (R^2 \nabla_i S / m_i) \tag{10.20}$$

implying that the quantum potential is *nonlocal*. In the case where the function $R(t, \vec{x}_1, ..., \vec{x}_n)$ is not a product of the form $R_1(t, \vec{x}_1)...R_n(t, \vec{x}_n)$ we have an *EPR* situation and Bell's inequalities are invalidated due to the nonlocal quantum force described by Q.

As Bohm said[1], the *whole* defines the behavior of the *parts*, or the parts behave in a special way, participating in the *holomovement*, as if taking part in some organized dance.

As in the one-particle case, Eq. (10.17) is interpreted as a generalized Hamilton-Jacobi equation for a system of n particles with a set of momentum fields

$$\vec{P}_i(t, \vec{x}_1, ..., \vec{x}_n) = \nabla_i S(t, \vec{x}_1, ..., \vec{x}_n). \tag{10.21}$$

The velocity fields are given by $\vec{V}_i = \vec{P}_i / m_i$ and the particle trajectories are the solutions $\vec{x}_i(\sigma_i(t)) \equiv \vec{x}_i(t)$ of the following system of differential equations $\vec{x}_j$

$$\frac{d}{dt}\vec{x}_i(t) = \frac{1}{m_i} \nabla_i S(t, \vec{x}_1, ..., \vec{x}_n)|_{\vec{x}=\vec{x}(\sigma_j(t))}, \quad i,j = 1, ..., n, \tag{10.22}$$

which can be solved once we know the initial positions $\vec{x}_{i0}$ of the particles.

10.3 REMARKS ON THE NON RELATIVISTIC BBT

(i) *Empty waves.* Suppose that the n-particle wave function is a superposition of two waves,

$$\Psi(t, \vec{x}_1, ..., \vec{x}_n) = \Psi_1(t, \vec{x}_1, ..., \vec{x}_n) + \Psi_2(t, \vec{x}_1, ..., \vec{x}_n). \tag{10.23}$$

The amplitude of this wave is

$$R^2 = R_1^2 + R_2^2 + 2R_1 R_2 \cos[(S_1 - S_2)/\hbar]. \tag{10.24}$$

Now, suppose that Ψ_1 and Ψ_2 have non common support in $\mathcal{C}_{3n}$. The system point is located in one of the regions of $\mathcal{C}_{3n}$, let us say, the one corresponding to Ψ_2. Then, Ψ_1 is said to be an *empty wave* and it has no influence on the motion of the particles, for under the above conditions Eq. (10.23) reduces to

$$R^2 \approx R_1^2 + R_2^2 \tag{10.25}$$

and Ψ becomes equivalent to a classical mixture. It is important to recall here that if subsequently Ψ_1 and Ψ_2 overlap, then Ψ_1 will actively influence the motion of the particles again. There are proposals in literature to observe the effects of empty waves.[3] However, Holland[2] showed that they cannot be observed, in the sense of producing results contradicting the predictions of standard quantum theory.

(ii) *Measuring process.* The main idea of the measurement theory in *BBT* is that it is a physical process like any other, i.e., in a first stage, the measuring device and the observed system interact. Then, the wave function of the combined system *bifurcates*, i.e., it breaks down into a sum of non-overlapping packets, each one corresponding to a possible result of the measurement. At a second stage, the detection device produces some magnification revealing the result obtained.

When the wave function of the system plus apparatus bifurcate, the observed system enters one of the branches, the others remaining empty, like in the case studied above. In this way the phenomenon of measurement looks like a wave packet collapse, but here, of course, there is no collapse and there is no need to appeal to any subjective process.

The measurement theory in *BBT* implies that quantum properties are not attributable, in general, to the observed system alone. This means that measurements are context-dependent and in general do not reveal the value of a physical property, *prior* to measurement, but give the result in which the observed system is left *after* its interaction with the measuring device. This point is crucial and Bohm and Hilley show that the mathematical formalism of the theory implies that with the exception of the position, all other properties of a system are not intrinsic, this meaning that the true value prior to measurement cannot be known.

(iii) *Calculation of averages.* What has been said in (ii) is well illustrated when we study the calculation of averages. There is a difference in the calculation of averages in *BBT* in relation to standard quantum theory. For example, the average of the momentum of a single particle is equal in *BBT* and standard quantum theory as is easy to verify, but for the average of the square of the momentum we have,

$$\langle \vec{p}^2 \rangle_{BBT} = \int d^3x \rho (\nabla S)^2 \qquad (10.26)$$

whereas, standard quantum theory gives for the average of the momentum (which is the correct experimental value)

$$\langle \vec{p}^2 \rangle = -\hbar^2 \int d^3x \Psi^* \nabla^2 \Psi = \int d^3x \rho [(\nabla S)^2 + \hbar^2 \frac{(\nabla R)^2}{R^2}]. \qquad (10.27)$$

These equations show that in general, the average value of an operator as computed in quantum theory cannot be obtained by averaging the corresponding physical property of the particle with the density ρ. Bohm and Hilley said that "we understand this as a consequence of the fact that the momentum is not an intrinsic property, but rather is inherently context dependent so that it involves the participation of the measuring apparatus and the quantum fields as a whole."

This last statement, according to us, does not justify the use of Eq. (10.27) in *BBT*.

Notes

1. We call the reader's attention to Holland[2] "if we wish to compare the methods, content, claims and experimental predictions of two physical theories, we have to find some common ground between them." The mathematical methods in which the de Broglie- Bohm theory is usually presented (and which is the one that we will use in what follows) does not adequately satisfy Holland's statement, because the relativistic and non-relativistic wave functions (and fields) and the classical description of relativistic and non-relativistic particle motion are based on objects of a very different mathematical nature. A presentation of all these theories using objects of the same mathematical nature (the so called Clifford and spin-Clifford bundle formalism[3−6]) reveals unsuspected relationships between the description of the motion of a classical charge in an electromagnetic field and the Dirac equation. Also, for the many-particle problem this formalism gives hints for an alternative formulation of the many-particle problem along lines suggested in[7−10]. Instead of having a wave function (see section 10.2) in configuration space we have n-fields in real spacetime. Some of these ideas are discussed in in [12].

2. Holland[2] writes energy-momentum conservation equations for Ψ using the standard Lagrangian formalism, but in this case the situation seems even worse than in the case of the Bohm and Hilley *suggestion*, because particle and wave "interact" in a non symmetric way. Ψ can transfer energy to the particle, but not vice-versa.

3. See [14] and references therein.

References

[1] D. Bohm and B. J. Hilley, *Undivided Universe* (Routledge, London and New York, 1993).

[2] P. Holland, *The Quantum Theory of Motion* (Cambridge University Press, Cambridge, 1993).

[3] D. Hestenes, in: *Clifford Algebras and Their Applications in Mathematical Physics*, edited by J. S. R. Chisholm and A. K. Common (D. Reidel Publishing Co., Dordrecht, 1986), pp.321–346.

[4] W. A. Rodrigues, Jr. and Q. A. G. de Souza, The Clifford bundle and the nature of the gravitational field, *Found. Phys.* **23**(11), 1465–1490 (1993)

[5] W. A. Rodrigues, Jr., Q. A. G. de Souza, J. Vaz, Jr. and P. Lounesto, Dirac-Hestenes spinor fields on Riemann-Cartan manifolds, *Int. J. Theor. Phys.* **35**(9), 1849–1900 (1996)

[6] W. A. Rodrigues, Jr., J. Vaz, Jr. and M. Pavsic The Clifford Bundle and the Dynamics of the Superparticle, *Banach Center Publ., Polish Acad. Sci.* **37**, 295–314 (1996).

[7] A. O. Barut, Schrödinger's interpretation of Ψ as a continuous charge distribution, *Annalen der Phys.* **45**(7), 31–36 (1998).

[8] A. O. Barut and N. Ünal, A new approach to bound-state quantum electrodynamics.1. Theory, *Physica A* **142**(1-3), 467–487 (1987).

[9] A. O. Barut and J. P. Dowling, Quantum electrodynamics based on self-energy without 2nd quantization—the Lamb shift and long range Casimir-Polder- van der Walls forces near boundaries, *Phys. Rev. A* **36**(6), 2250–2556 (1987).

[10] A. O. Barut, The revival of Schrödinger's interpretation of quantum mechanics, *Found. Phys. Lett.* **1**(1), 47–56 (1988).

[11] L. de Broglie, *Non Linear Quantum Mechanics. A Causal Interpretation* (Elsevier Publishing Co., Amsterdam, 1960).

[12] W. A. Rodrigues, Jr., $0 \leq v < \infty$ solutions of the relativistic wave equations and the foundations of relativity and quantum theory, RP 36/98, IMECC-UNICAMP, (1998).

[13] R. Gutler and D. Hestenes, Consistency in the formulation of the Dirac, Pauli, and Schrödinger theories, *J. Math. Phys.* **16**(3), 573–583 (1973).

[14] F. Selleri, *Quantum Paradoxes and Physical Reality* (Kluwer Academic Publishers, Dordrecht, 1990).

11
DE BROGLIE-BOHM RELATIVISTIC HVT

In this Chapter $\mathcal{M} = (M, D, \eta)$, denotes Minkowski spacetime[1], i.e., M is a four dimensional manifold diffeomorphic to $\mathcal{R}^4$, η is a Lorentzian metric of signature $(1, 3)$ and D is the Levi-Civita connection of η. As is well known, under these conditions there exists a global coordinate chart $\langle x^\mu \rangle$ of the maximal atlas of M, such that there is a global basis $\{\partial/\partial x^\mu\}$ of TM (the tangent bundle) such that[1]

$$\eta_{\mu\nu} = \eta(\partial/\partial x^\mu, \partial/\partial x^\nu) = diag(1, -1, -1, -1). \tag{11.1}$$

11.1 HVT INTERPRETATION OF THE BOSON EQUATIONS

Let $\psi : M \to \mathcal{C}$ be the Klein-Gordon-Fock (*KGF*) wave function and $A \in \sec T^\star M$ be the external electromagnetic potential[2]. Then, the *KGF* equation for a particle of mass m is (in the chart $\langle x^\mu \rangle$)

$$(i\partial_\mu - eA_\mu)(i\partial^\mu - eA^\mu)\psi = m^2\psi. \tag{11.2}$$

If we substitute

$$\psi = R\exp(iS/\hbar) \tag{11.3}$$

into Eq. (11.2), we get

$$(\partial_\mu S + eA_\mu)(\partial^\mu S + eA^\mu) = m^2(1 + Q) \tag{11.4}$$

$$\partial_\mu j^\mu = 0 \tag{11.5}$$

where the KGF current is given by

$$j_\mu = \frac{1}{2m}\psi^*(i\overleftrightarrow{\partial_\mu} - 2eA_\mu)\psi = -\frac{R^2}{m}(\partial_\mu S + eA_\mu). \quad (11.6)$$

Equation (11.4) looks like the relativistic Hamilton-Jacobi equation for a particle with mass $m' = m(1+Q)^{\frac{1}{2}}$, where Q, the quantum potential is given by[3]

$$Q = \frac{\Box R}{m^2 R}. \quad (11.7)$$

Despite the existence of a conserved current we cannot use it to define a guidance law, because j_0 is not positive definite and worse, j_μ is not in general a time-like vector. If, e.g., we define

$$m'u_\mu = p_\mu = -(\partial_\mu S + eA_\mu), \quad (11.8)$$

then vector u^μ cannot, in general, have the meaning of a velocity vector, because the variable mass m' can be imaginary. Accepting Eq. (11.8) would imply in situations for which there are solutions of the *KGF* equation for which the particle trajectory which start as a timelike trajectory becomes spacelike and can even move backwards in time. This is certainly unacceptable, and leads us to affirm that the *KGF* equation does not have a consistent single particle interpretation.

We can verify that analogous problems exist for all boson equations, but as we shall see in the next section, these problems do not exist for the fermion wave equation (Dirac equation). In particular, the interpretation of Maxwell equations in the *BBT* does not allow for the existence of *photons* as beables[2,3].

The maximum that can be done according to Holland[3], is to associate with the boson equations an energy flow based on the standard energy-momentum tensor that follows from Lagrangian field theory. Without entering into details here, we would like to comment that this method can fail in some situations. Indeed, recent experiments show that for sound waves there can exist energy flows that do not travel according to the standard way suggested by the theory of the energy-momentum tensor. This happens for some extraordinary field configurations that can travel in an homogeneous medium with speed greater then c_s, the so called sound speed parameter that appears in the homogeneous wave equation (for details see[4]).

Due to the problems quoted above, Bohm and Hilley[2] proposed to use a completely different theory for bosons: instead of particles, some *random classical fields* are taken as *beables*. Surely, this different treatment of fermions and bosons, which in the classical limit gives rise to classically moving massive particles and form the classical world, due to the possibility of having a large amount of bosons in the same state, looks *artificial*, to say the least.

The idea of the quantum potential for bosons leads to a change of the respective classical wave equations. This happens because the boson field is *guided* by the boson wave function. For example, for massless spinless field, instead of the classical wave equation,

$$\Box \phi = 0 \quad (11.9)$$

we get,

$$\Box\phi + \frac{\delta Q}{\delta\phi} = 0 \qquad (11.10)$$

where Q, now called the *super quantum potential*, is given by

$$Q = -\frac{1}{2}\int_{\text{all space}} dV \left\{ \frac{[\delta^2/(\delta\phi(x^\mu))^2 R(...\phi(x^\mu)...)]}{R(...\phi(x^\mu)...)} \right\} \qquad (11.11)$$

where in analogy with the non-relativistic particle theory, the wave function is now a wave functional that can be written

$$\Psi(...\phi(x^\mu)...) = R(...\phi(x^\mu)...)\exp(iS(...\phi(x^\mu)...)/\hbar) \qquad (11.12)$$

The wave functional Ψ depends on the whole set of values of ϕ at every point $x \in M$, and solves the super Schrödinger equation,

$$i\hbar\frac{\partial\Psi}{\partial t} = \mathbf{H}\Psi \qquad (11.13)$$

where $\mathbf{H}$ is the super Hamiltonian, given by

$$\mathbf{H} = \frac{1}{2}\int_{\text{all space}} dV \left[-\frac{\delta^2}{(\delta\phi(t,\vec{x}))} + (\nabla\phi(t,\vec{x}))^2 \right]. \qquad (11.14)$$

The resulting theory is a non-relativistic one, because the nonlocality of the quantum potential implies in a breakdown of Lorentz invariance, which may result in phenomena not in accord with current scientific ideas.

11.2 THE BBT INTERPRETATION OF THE DIRAC EQUATION

Let $\psi : M \to \mathcal{C}^4$ be a Dirac spinor field[5] for a particle of mass m. Then the Dirac equation for the particle in interaction with an external electromagnetic field is,

$$i\gamma^\mu(\hbar\partial_\mu - ieA_\mu)\psi = 0 \qquad (11.15)$$

where $\{\gamma^\mu\}$, $\mu = 0, 1, 2, 3$ are a set of Dirac matrices.

The Dirac equation implies in the existence of a conserved time-like current,

$$j^\mu = \overline{\psi}\gamma^\mu\psi \qquad (11.16)$$
$$\partial_\mu j^\mu = 0 \qquad (11.17)$$

where $\overline{\psi} = \psi^\dagger\gamma^0$, the symbol † denoting the Hermitian conjugate. From this current, it is possible to define a (time-like) velocity field V^μ, from which a guidance law follows. Calling,

$$a = \overline{\psi}\psi, \quad b = i\overline{\psi}\gamma^\mu\gamma^5\psi \qquad (11.18)$$

where $\gamma^5 = \gamma^0\gamma^1\gamma^2\gamma^3$, we have,

$$V^\mu = \frac{j^\mu}{\sqrt{a^2+b^2}} \qquad (11.19)$$

As in the non-relativistic *BBT*, here also the particle is supposed to follow a time like world line, which is one of the integral lines of the vector field V^μ. The worldlines $\{\sigma\}$ of the ensemble of particles associated with a solution of Eq. (11.15) are then the solutions (in a given Lorentz chart $\langle x^\mu \rangle$) of the differential equation

$$v^i(t) = \frac{d}{dt}x^i(\sigma(t)) = \frac{j^i}{j^0}|_\sigma. \qquad (11.20)$$

This equation gives plausible results for the tunneling time of electrons through a potential barrier[3,6].

It is interesting to recall that the mean value of the velocity in the ensemble coincides with the usual quantum theory result,

$$\langle v^i \rangle = \int d^3x\, j^0 v^i = \int d^3x\, \overline{\psi}\gamma^i\psi = \langle \gamma^0\gamma^i \rangle. \qquad (11.21)$$

The eigenvalues of $\gamma^0\gamma^i$ are ± 1, while v^i is a continuous function. This strange feature is associated with the so called zitterbewegung motion. We are not going to discuss this topic here. The interested reader may consult [7-9]. Another point that is worth mentioning is that in Dirac theory the momentum density (obtained from the usual Lagrangian formalism) is not, in general, parallel to the velocity field, which means that in the *BBT* interpretation of the Dirac equation, the flux of momentum and energy are not directly associated with the translational motion of the spinning particle. This is a very important issue, but it will not be discussed here. We only mention that the theory has a natural 3-spin vector per particle,

$$w^i = \frac{1}{2}\frac{\overline{\psi}\gamma^i\gamma^5\psi}{j^0}, \qquad (11.22)$$

from which a very natural explanation for the interactions of spin 1/2 particles in a Stern-Gerlach experiment results (see [3,6,10]).

11.3 RELATIVISTIC BBT MANY-FERMION THEORY

The Dirac wave function for an n-particle fermion system of masses m_j, $j = 1, ..., n$ is the mapping

$$\begin{aligned}
\Psi \;:\; & \mathcal{R}\times\mathcal{C}_{3n} \to \mathcal{C}^4 \otimes .... \otimes \mathcal{C}^4 \\
(t,\vec{x}_1,...,\vec{x}_n) \;\mapsto\; & \sum c_{a_1a_2...a_n}\psi_{a_1}(t,\vec{x}_1)\otimes...\otimes\psi_{a_n}(t,\vec{x}_n) \\
= & \sum c_{a_1a_2...a_n}\Psi_{a_1...a_n}(t,\vec{x}_1,...,\vec{x}_n). \qquad (11.23)
\end{aligned}$$

Ψ, (which is appropriately antisymmetric in the indices $a_1,..a_n$) satisfies the following n-particle Dirac equation when in interaction with the electromagnetic field[4],

$$i\hbar \frac{\partial \Psi}{\partial t} = \sum_{j=1}^{n}\sum_{p=1}^{3} -i\left\{\left[\hbar \alpha_p^{(j)}(\nabla_p^{(j)} - \vec{A}_p^{(j)}) + m_j \beta^{(j)} + V^{(j)}\right]\right\}\Psi \qquad (11.24)$$

where

$$\begin{aligned}\alpha_p^{(j)} &= \mathbf{1} \otimes ... \gamma_0^{(j)}\gamma_p^{(j)} \otimes ... \otimes \mathbf{1} \\ \beta^{(j)} &= \mathbf{1} \otimes ... \gamma_0^{(j)} \otimes ... \otimes \mathbf{1}\end{aligned} \qquad (11.25)$$

and the matrices $\gamma_\mu^{(j)}, \gamma_\nu^{(k)}, \mu,\nu = 0,1,2,3, j,k = 1,...,n$ satisfy

$$\begin{aligned}\{\gamma_\mu^{(j)},\gamma_\nu^{(k)}\} &= 2\eta_{\mu\nu}\delta^{jk} \\ [\gamma_\mu^{(j)},\gamma_\nu^{(k)}] &= 0, \text{ if } j \neq k\end{aligned} \qquad (11.26)$$

From Eq. (11.24) we immediately get,

$$\frac{\partial \Psi^\dagger \Psi}{\partial t} + \sum_{j=1}^{n}\sum_{p=1}^{3} \nabla_p^{(j)} \Psi^\dagger \alpha_p^{(j)} \Psi = 0 \qquad (11.27)$$

Then, putting $\rho = |\Psi|^2$ we have the guidance law for the jth particle,

$$\vec{v}^{(j)} = \frac{\Psi^\dagger \alpha^{(j)} \Psi}{\rho} \qquad (11.28)$$

and the following conservation equation in configuration space

$$\frac{\partial \rho}{\partial t} + \sum \nabla^{(j)} \cdot (\rho \vec{v}^{(j)}). = 0 \qquad (11.29)$$

11.4 NECESSITY OF SUPERLUMINAL PROCESSES

The formulation of the *BBT* n-particle theory needs the fixation of a preferred Lorentz frame and implies in a breakdown of Lorentz invariance. This can be seen at once if we examine, e.g., the case of a two-particle system, for which Ψ is a superposition,

$$\Psi = A\psi_a(t,\vec{x}_1) \otimes \psi_b(t,\vec{x}_2) + B\psi_c(t,\vec{x}_1) \otimes \psi_d(t,\vec{x}_2). \qquad (11.30)$$

First we observe that there is a nonlocal connection between the particles, since the probability density in configuration space is

$$\begin{aligned}|\Psi|^2 &= |A|^2|\psi_a(t,\vec{x}_1)|^2|\psi_b(t,\vec{x}_2)|^2 \\ &+ |B|^2|\psi_c(t,\vec{x}_1)|^2|\psi_d(t,\vec{x}_2)|^2 \\ &+ A^*B\psi_a^\dagger(t,\vec{x}_1)\psi_b^\dagger(t,\vec{x}_2)\psi_c(t,\vec{x}_1)\psi_d(t,\vec{x}_2) + c.c.\end{aligned} \qquad (11.31)$$

The current for the first particle is

$$\begin{aligned}\vec{j}^{(1)} &= |A|^2 \psi_a^\dagger(t,\vec{x}_1)\alpha^{(1)}\psi_a^\dagger(t,\vec{x}_1)|\psi_b(t,\vec{x}_2)|^2 \\ &+ |B|^2 \psi_c^\dagger(t,\vec{x}_1)\alpha^{(2)}\psi_c(t,\vec{x}_1)|\psi_d(t,\vec{x}_2)|^2 \\ &+ A^*B\psi_a^\dagger(t,\vec{x}_1)\alpha^{(1)}\psi_c(t,\vec{x}_1)\psi_b^\dagger(t,\vec{x}_2)\alpha^{(2)}\psi_d(t,\vec{x}_2) \\ &+ c.c. \end{aligned} \quad (11.32)$$

For the second particle, we have a similar expression with $1 \leftrightarrow 2$.

From these equations we see that the velocities of the particles $\vec{v}^{(i)} = \vec{j}^{(i)}/\rho$ are not independent. The velocity of each particle will depend on the velocity of the other particle for situations in which the spacetime interval between the two particles is spacelike. This instantaneous (superluminal processes) influence chooses a preferred Lorentz frame and implies in a breakdown of Lorentz invariance. Only when the nonlocal connections can be neglected is it that we have Lorentz invariant processes.

In standard quantum theory, we have what can be called *passive nonlocality*. There are random processes in two points in spacetime separated by a space-like interval (unpredictable in principle). It occurs that when comparing the results of observations, we find correlations for these random processes— this is *passive nonlocality*. We cannot use this correlation to send any superluminal signal containing any information (unless, as observed in Chapter 4, there exists a psychokinetic effect[11–14]). The ultimate quantum indeterminism for individual observations implies that there is no chance for us to dominate these random processes. But, if there is an ultimate deterministic theory for quantum phenomena, like the one described above, then, as it is the case of classical physics, a nonlocal correlation must be described by some superluminal field (or particle), which of course will breakdown Einstein's causality and the whole structure of the orthodox interpretation of special relativity based on Minkowski spacetime. This we call *active nonlocality*. If active nonlocality exists or not in the physical world where we live, is an open question. Indeed, we already called the reader's attention (footnote no. 3 of Chapter 4) to the fact that all the relativistic equations have solutions traveling with arbitrary speed $0 \leq v < \infty$. Of course, if we can build,[5] let us say an electromagnetic field configuration which constitutes a signal and that travels with $v \geq c(=1)$ we will have a breakdown of the principle of relativity, since as proved in[4], Lorentz invariance will be broken down in an active and passive senses, for otherwise we will generate unsurmountable logical paradoxes for any Boolean mind.

We end this section with the very important observation that the statistical predictions of the n-particle *BBT* described above are Lorentz invariant. The interested reader may find the proof in[2].

In the last three Chapters, we discussed and criticized the ideas related to hidden variables theories as providing what Bohm and Hilley call a *realistic or ontological interpretation* of the quantum theory (*RIQT*)[6]. We must say that given the actual status of the *BBT* theory,"ontological interpretation" or "realistic interpretation" are misnomers, for certainly it is a difficult task to say that we are using good sense[7] and at the same time considering as real, a wave that is propagating in configuration space[8]. However this is exactly what Bohm & Hiley, Holland and their fellows do. Holland[3], e.g., explicitly states: "*we ascribe to configuration space as much physical reality as we do to three-dimensional Euclidean space in the one body theory. This is a radical step, but a necessary one if we wish to extend the quantum theory of motion consistently to embrace many body problems."*

It must be said that in de Broglie's original attempts[15] to develop a realistic interpretation, he devoted a great effort in trying to formulate the concept of the wave function of a many particle system simply as a convenient description of a beable called the non linear field and which lives and moves in spacetime (and whose singularities are the particles[9]). However, he did not succeed in providing a consistent theory, and certainly his fellows did not obtain any real improvement[10]. This, of course, does not mean that such a theory does not exist. Even more, a theory of this kind may suggest to us some crucial experiment for which quantum theory makes a different prediction and may eventually lead us to find the limit of validity of current quantum theory. An open mind is worthwhile here if we wish to expand our knowledge.

Notes

1. In this section we use units such that the speed of light $c = 1$.

2. $\sec T^* M$ means section of the cotangent bundle.

3. In Eq. (11.7), $\Box = \partial^2/\partial t^2 - \nabla^2$ is the wave operator.

4. Here the electromagnetic field is the one generated by the particles. If necessary, an eventual external field can be added.

5. There is some experimental evidence[25,26] that superluminal electromagnetic X-waves can be launched in physical space, but the authors of[25] are of the opinion that this superluminal wave is not a signal.

6. Holland prudently prefers to call the developments of these ideas "The Quantum Theory of Motion", this being the title of his recent book[3].

7. The only thing that according to Voltaire all men think they have enough of...

8. Well, this conclusion certainly will not be admitted by a Platonist, since he will say that Hilbert spaces, configuration spaces and their elements, etc. are real....

9. Eventually, today he would call the particles soliton solutions of the non-linear wave equation, which he would never manage to find. Vigier, one of de Broglie's fellows in developing ideas of Mackinon[16], proposed a non linear wave equation for a spin zero boson field[17-18]. It is interesting to point out here that there are undistorted progressive waves (*UPWs*), solutions for both the scalar homogeneous wave equation (*HWE*) and also for Klein-Gordon-Fock equation (*KGFE*), that are equal to the soliton of the nonlinear equation studied by Vigier. Even more so, there are an infinity of *UPW* solutions of these equations (and also of the other relativistic wave equations) that move with arbitrary speed $0 \leq v < \infty$. See[4] for details.

10. Besides the earlier attempts [19], there are some fresh ideas concerning this issue in[21-24]

References

[1] R. K. Sachs and H. Wu, *General Relativity for Mathematicians* (Springer, New York, 1997).

[2] D. Bohm and B. J. Hilley, *Undivided Universe* (Routledge, London and New York, 1993).

[3] P. R. Holland, *The Quantum Theory of Motion* (Cambridge University Press, Cambridge, 1993).

[4] W. A. Rodrigues, Jr. and J. Y. Lu, On the existence of undistorted progressive waves (UPWs) of arbitrary speeds $0 \leq v < \infty$ in nature, *Found. Phys.* **27**(3) 453–508 (1997).

[5] W. A. Rodrigues, Jr. , Q. A. G. de Souza, J. Vaz, Jr. and P. Lounesto, Dirac-Hestenes spinor fields in Riemann-Cartan manifolds, *Int. J. Theor. Phys.* **35**(9), 1849–1900 (1996).

[6] A. Lasenby, S. Gull and C. Doran, in: *Clifford (Geometrical) Algebras with Applications in Physics, Mathematics and Engineering*, edited by W. E. Baylis (Birkhäuser, Boston, 1996), pp.147–169.

[7] W. A. Rodrigues, Jr., J. Vaz, Jr, E. Recami and G. Sallesi,, About zitterbewegung and electron structure, *Phys. Lett. B* **318**(4), 623–628 (1993).

[8] W. A. Rodrigues, Jr. and J. Vaz, Jr., in: *Electron Theory and Quantum Electrodynamics*, edited by J. P. Dowling (Plenum Publ. Corporation, New York, 1997), pp. 201–222.

[9] W. A. Rodrigues, Jr., J. Vaz, Jr. and M. Pavsic The Clifford bundle and the dynamics of the superparticle, *Banach Center Publ., Polish Acad. Sci.* **37**, 295–314 (1996).

[10] W. E. Baylis, , in: *Clifford (Geometrical) Algebras with Applications in Physics, Mathematics and Engineering*, edited by W. E. Baylis (Birkhäuser, Boston, 1996), pp.253–268.

[11] H. Schmidt, Collapse of the state vector and PK effect, *Found. Phys.* **12**(6) 565–581 (1982).

[12] H. P. Stapp, Theoretical model of a purported empirical violation of the predictions of quantum theory, *Phys. Rev. A* **50**(1), 18–22 (1994).

[13] E. J. Squires, An experiment to test an explanation of a possible violation of quantum theory, *Found. Phys. Lett.* **8**(6), 589–591 (1995).

[14] O. Costa de Beauregard, Macroscopic retrocausation, *Found. Phys. Lett.* **8**(3), 287–291 (1995).

[15] L. de Broglie, *Non Linear Quantum Mechanics. A Causal Interpretation*, Elsevier Publishing Co., Amsterdam, 1960.

[16] L. Mackinnon, Nondispersive de Broglie wave packet, *Found.Phys.* **8**(3-4), 157–176 (1978).

[17] Ph. Gueret and J. P. Vigier, De Broglie wave particle duality in the stochastic interpretation of quantum mechanics—a testable physical assumption, *Found. Phys.* **12**(11), 1057–1083 (1982);

[18] Ph. Gueret and J. P. Vigier, Relativistic wave equations with quantum potential, **38**(4), 125–128 (1983).

[19] H. Freidstadt, The causal formulation of quantum mechanics of particles, *Nuovo Cimento* **5**(suppl.), 1–70 (1957).

[20] A. O. Barut, Schrödinger's interpretation of Ψ as a continuous charge distribution, *Annalen der Phys.* **45**(7), 31–36 (1998).

[21] A. O. Barut and N. Ünal, A new approach to bound-state quantum electrodynamics. 1. Theory, *Physica A* **142**(1-3), 467–487 (1987).

[22] A. O. Barut and J. P. Dowling, Quantum electrodynamics based on self-energy without 2nd quantization—the Lamb shift and long range Casimir-Polder- van der Walls forces near boundaries *Phys. Rev. A* **36**(6), 2250–2556 (1987).

[23] A. O. Barut, The revival of Schrödinger's interpretation of quantum mechanics, *Found. Phys. Lett.* **1**(1), 47–56 (1988).

[24] P. Saari and K. Reivelt, Evidence for X-shaped propagation-invariant localized light waves, *Phys. Rev. Lett.* **79**(21), 4135–4138 (1997).

[25] E. Capelas de Oliveira and W. A. Rodrigues Jr., Superluminal electromagnetic waves in free space, *Ann. der Physik* **7**(7-8), 654–659 (1998).

12
STATISTICAL INTERPRETATION

12.1 INTRODUCTION

Here we discuss some aspects of the so called *statistical interpretation* of quantum theory (*SIQT*), which rejects the idea that the wave function describes one individual quantum system. The main claim of the *SIQT*, which originated from some of Einstein's ideas[1], is that a quantum state, which may be *pure* and then described by a state vector $|\Psi\rangle$ or *mixed* and in this case described by a density matrix ρ, represents an *ensemble* of identically prepared systems. We shall denote the ensemble by $\mathcal{E}_{|\Psi\rangle}$ or $\mathcal{E}_\rho$ depending on the case.

If the quantum system is, e.g., a hydrogen atom, then the ensemble $\mathcal{E}_{|\Psi\rangle}$ will be the conceptual infinite set of single hydrogen atoms that have been prepared in the same quantum state $|\Psi\rangle$ by an appropriate preparation technique.

For our example where the quantum system is a hydrogen atom, it is very important in *SIQT* not to confuse the ensemble with a beam of hydrogen atoms, which is a many particle system. A beam of hydrogen atoms such that its intensity is low and only one atom is presented at a time, simulates an ensemble of a single hydrogen atom[2]. For *SIQT*, it is important also to take into account that the quantum ensembles are conceptually different from the ones used in statistical thermodynamics, where, for calculations, we use a representative ensemble, but the results of our calculations can be compared with a measurement in a single system. The reason for this distinction is that *SIQT* claims that quantum theory predicts nothing that is relevant to a single measurement in a single system and then the results of our calculations apply only to an ensemble of similar measurements.

133

If *SIQT* is the right interpretation of quantum formalism we can imagine, as we already said in Chapter 9, that individual particles can be described by something else, e.g., by some hidden variables giving a more profound description of them. Arguing against *SIQT* in discussions with Blokhintsev, Fock[3] noticed, that if we did not change the formalism of quantum theory, then, since in experiments with a *quantum ensemble*, the particles that are prepared identically do not interact with one another, for them there is no observable which can discriminate the description of one individual particle from the description of the ensemble. This, according to Fock, is the reason if we suppose the completeness of quantum formalism, we must consider the state vector to be the complete and ultimate description both of the individual particle as well as of the quantum ensemble.

Another argument against *SIQT* is that the idea that the state vector describes an individual object is fruitful, since we can predict properties of this object using ideas of symmetry. As an example, consider the theory of elementary particles, where Gell-Mann using $SU(3)$ symmetry, and supposing that the state vectors of the "fundamental" elementary particles (quarks) are members of the carrier space of the irreducible representation of this group, predicts the properties of Ω^- particles.

Recently also, some authors based on their experimentation claim to have observed individual atoms and quantum processes [5,6] which confirm the standard quantum theory. In [5], an atom in "a Schrödinger's cat state" superposition is observed!

Nevertheless, if we generalize the formalism of quantum theory, in such a way that the individual particles are described besides $|\Psi\rangle$, also by some hidden variables, then in order to see what kind of experiments are necessary to reveal their existence, we must look at what the special features are that characterize such a theory. So, let us suppose (that contrary to the wisdom of the orthodox interpretation of quantum formalism), an individual particle *has* as *beables,* properties which are described by noncommuting operators. During a measurement in the ensemble of particles, by *having* these characteristics, some *subensemble* is selected, where the observer sees one of these properties.

Let $|\Psi\rangle$ be the state vector of our statistical ensemble $\mathcal{E}_{|\Psi\rangle}$ of equally prepared systems, each of which *has* the same values for all dynamical variables. The values of these variables are distributed in such a way that the probability that in a random observation for some element in the ensemble, observable A (represented by the Hermitian operator **A**) has the value a_i, is given by the quantum theory rule

$$w_i = |\langle u_i|\Psi\rangle|^2, \quad \mathbf{A}|u_i\rangle = a_i|u_i\rangle \qquad (12.1)$$

Also, in the *SIQT,* if two complementary observables A, B (represented by the Hermitian operators **A**,**B**) are measured in the ensemble, then the dispersions (standard deviations) $\triangle \mathbf{A}$ and $\triangle \mathbf{B}$ satisfy the Heisenberg uncertainty relations. However, in the *SIQT*, Heisenberg uncertainty relations have nothing to do with limitations imposed on measurements. A good discussion of this issue can be found in [2].

12.2 WHEN IS THE SIQT POSSIBLE?

Let us discuss now the example, given in [7], which shows in what cases the statistical interpretation is possible. In Chapter 4 we discussed Bell's inequalities in d'Espagnat form for three observables A, B, C. We found (see Eq. (4.4)) that the number of systems in the ensemble with the supposed properties satisfies

$$N(A^+B^-) \leq N(B^-C^+) + N(A^+C^-). \tag{12.2}$$

There, in fact, by considering pairs of correlate particles, we argued against the *SIQT* taking into account standard probability calculus and validity of special relativity. However, let us investigate the issue (for a spin 1/2 system) more carefully, following [7,8].

At first, suppose that our system is in the state $|\psi\rangle$ and that two observables A, B, represented by the noncommuting operators **A** and **B** are measured. As eigenvectors of these operators we have two orthogonal vectors on the plane

$$\mathbf{A}|a_i\rangle = a_i|a_i\rangle$$
$$\mathbf{B}|b_i\rangle = b_i|b_i\rangle \tag{12.3}$$

Let $\phi = \theta_a - \theta_e$ (see Fig.(12.1)) be the angle between the vectors $|a_+\rangle, |b_+\rangle$ on a two-dimensional plane representing the Hilbert space of a single 1/2 particle.

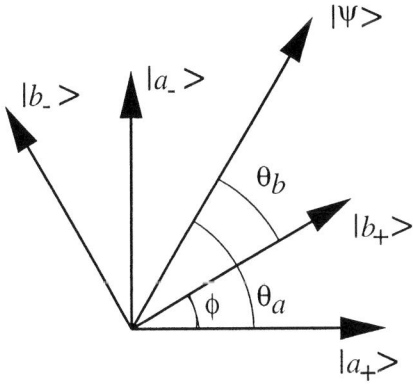

Figure 12.1. Diagram for the example of section 12.2.

Now suppose that $0 < \phi < \frac{\pi}{2}, 0 \leq \theta \leq \pi$ and let us write

$$|\psi\rangle = \cos\theta_a |a_+\rangle + \sin\theta_a |a_-\rangle \qquad (12.4)$$
$$|\psi\rangle = \cos\theta_b |b_+\rangle + \sin\theta_b |b_-\rangle \qquad (12.5)$$
$$|b_+\rangle = \cos\phi |a_+\rangle + \sin\phi |a_-\rangle \qquad (12.6)$$
$$|b_-\rangle = -\sin\phi |b_+\rangle + \cos\phi |b_-\rangle \qquad (12.7)$$
$$|a_+\rangle = \cos\phi |b_+\rangle - \sin\phi |b_-\rangle \qquad (12.8)$$
$$|a_-\rangle = -\sin\phi |b_+\rangle + \cos\phi |b_-\rangle. \qquad (12.9)$$

Then the probabilities of having $A = a_+, A = a_-, B = b_+, B = a_-$ as results of measurements of A, B, are

$$p(A = a_+) = \cos^2\theta_a \qquad (12.10)$$
$$p(A = a_-) = \sin^2\theta_a \qquad (12.11)$$
$$p(B = b_+) = \cos^2\theta_b \qquad (12.12)$$
$$p(B = a_-) = \sin^2\theta_b. \qquad (12.13)$$

If the system is initially in the state $|\psi\rangle$ and after measuring A we measure B and vice versa, then the probabilities of definite results in such measurements are

$$p(A = a_+, \text{ then } B = b_-) = \cos^2\theta_a \sin^2\phi \qquad (12.14)$$
$$p(B = b_-, \text{ then } A = a_+) = \sin^2\theta_b \sin^2\phi. \qquad (12.15)$$

In the *SIQT*, the vector $|\psi\rangle$ characterizes the ensemble of systems $\mathcal{E}_{|\psi\rangle}$, in which each member has exact values $A(a_+ \text{ or } a_-)$ and $B(b_+ \text{ or } b_-)$. Introduce now the functions $f_{AB}(a_i, b_j)$, representing the subensemble of systems in $E_{|\psi\rangle}$ for which

$$A = a_i, \ B = b_j, \ i, j = \pm \qquad (12.16)$$

It is evident that $f_{AB}(a_i, b_j) \geq 0, (i, j = \pm)$ and

$$f_{AB}(a_+, b_+) + f_{AB}(a_+, b_-) = P(A = a_+) \qquad (12.17)$$

then from Eqs. (12.10)-(12.13):

$$f_{AB}(a_+, b_+) + f_{AB}(a_+, b_-) = P(A = a_+) = \cos^2\theta_a \qquad (12.18)$$
$$f_{AB}(a_-, b_+) + f_{AB}(a_-, b_-) = P(A = a_-) = \sin^2\theta_a \qquad (12.19)$$
$$f_{AB}(a_+, b_+) + f_{AB}(a_-, b_+) = P(B = b_+) = \cos^2\theta_b \qquad (12.20)$$
$$f_{AB}(a_+, b_-) + f_{AB}(a_-, b_-) = P(B = b_-) = \sin^2\theta_b. \qquad (12.21)$$

We have then, four unknown values $f_{AB}(a_i, b_j)$. However, Eq. (12.11) can be obtained as the sum of the first two equations minus the third, so that really we have three equations for four unknowns. As a solution of (12.11)-(12.21), we can take

$$f_{AB}(a_+, b_+) = \cos^2 \theta_a \cos^2 \theta_b$$
$$f_{AB}(a_+, b_-) = \cos^2 \theta_a \sin^2 \theta_b$$
$$f_{AB}(a_-, b_+) = \sin^2 \theta_a \cos^2 \theta_b$$
$$f_{AB}(a_-, b_-) = \sin^2 \theta_a \sin^2 \theta_b. \tag{12.22}$$

There are, however, other solutions. We could interpret each $f_{AB}(a_i, b_j)$ as the probability that a measurement of A and *then* B gives results (a_i, b_j). But from the definition of $f_{AB}(a_i, b_j)$ as that part of systems in the ensemble $\mathcal{E}_{|\Psi\rangle}$ which have (a_i, b_j), it follows that $f_{AB}(a_i, b_j) = f_{AB}(b_j, a_i)$ which contradicts quantum theory due to Eq. (12.14) and Eq. (12.15).

So, if we want to be free from contradiction and still believe into the *existence* of A, B together, we must have a *many-valuedness description*, i.e., different $f_{AB}(a_i, b_j)$ must be used for different *orders* of observations of A and B. Really, this means that there is a dependence of the system on the will of the observer, and the system must be described simultaneously by many (two) probability measures and frequencies.

But even such a sophisticated model becomes impossible, as we shall see now, if three noncommuting observables are measured.

12.3 SIQT × ORTHODOX QUANTUM THEORY

Consider a system $\mathfrak{S}$, composed of two subsystems $\mathfrak{S}_1, \mathfrak{S}_2$, each one being described by a two dimensional Hilbert space, respectively $\mathcal{H}^1$ and $\mathcal{H}^2$. Denote by $O_\mu, \mu = 1, 2$ the observables of the subsystem $\mathfrak{S}_\mu$, and by $\mathbf{O}_\mu$ the corresponding operators. Let $o_{\mu+}, o_{\mu-}$ and $|o_{\mu+}\rangle, |o_{\mu-}\rangle$ be respectively their eigenvalues and eigenvectors. Then, the vector $|\psi\rangle \in \mathcal{H}^1 \otimes \mathcal{H}^2$ describing our system $\mathfrak{S}$ can be written as

$$|\psi\rangle = k_1|o_{1+}\rangle_1 \otimes |o_{2+}\rangle_2 + k_2|o_{1+}\rangle_1 \otimes |o_{2-}\rangle_2 +$$
$$+ k_3|o_{1-}\rangle_1 \otimes |o_{2+}\rangle_2 + k_4|o_{1-}\rangle_1 \otimes |o_{2-}\rangle_2. \tag{12.23}$$

Then, for the probabilities quantum theory gives

$$P(O_1 = o_{1+}, O_2 = o_{2+}) = k_1^2 \tag{12.24}$$
$$P(O_1 = o_{1+}, O_2 = o_{2-}) = k_2^2 \tag{12.25}$$
$$P(O_1 = o_{1-}, O_2 = o_{2+}) = k_3^2 \tag{12.26}$$
$$P(O_1 = o_{1-}, O_2 = o_{2-}) = k_4^2. \tag{12.27}$$

As observables $\mathbf{O}_\mu$ take A_1, B_1, C_1 for $\mathfrak{S}_1$ and A_2, B_2, C_2 for $\mathfrak{S}_2$. Consider the basis which diagonalize $\mathbf{B}_\mu$ rotated relatively to the basis that diagonalize $\mathbf{A}_\mu$ by angles ϕ and 2ϕ, and take

$$0 \leq \phi < \frac{\pi}{4}. \tag{12.28}$$

Then,

$$|a_+\rangle_\mu = \cos\phi|b_+\rangle_\mu - \sin\phi|b_-\rangle_\mu \tag{12.29}$$
$$|a_-\rangle_\mu = \sin\phi|b_+\rangle_\mu + \cos\phi|b_-\rangle_\mu \tag{12.30}$$
$$|a_+\rangle_\mu = \cos 2\phi|c_+\rangle_\mu - \sin 2\phi|c_-\rangle_\mu \tag{12.31}$$
$$|a_-\rangle_\mu = \sin 2\phi|c_+\rangle_\mu + \cos 2\phi|c_-\rangle_\mu \tag{12.32}$$

Each vector $|\psi\rangle$ can be written as (12.23) where as the $\{|o_1\rangle\}$- basis we can take any one of the bases corresponding to the diagonalization of A_1, B_1, C_1, and as the $\{|o_2\rangle\}$-basis any one corresponding to the the diagonalization of A_2, B_2, C_2. So, there are 9 different possibilities for writing Eq. (12.23).

Let us think, following the *SIQT*, that there is an ensemble $\mathcal{E}_{|\Psi\rangle}$, systems with exact values for the 6 observables $A_1, B_1, C_1, A_2, B_2, C_2$. Let the function $f(a_i, b_j, c_k, ; a_l, b_m, c_n$ be the fraction of systems in the $\mathcal{E}_{|\Psi\rangle}$ ensemble, having $A_1 = a_i, B_2 = b_j, C_1 = c_k$, $A_2 = a_l, B_2 = b_m, C_2 = c_n$

Supposing that

$$f(a_i, b_j, c_k; a_l, b_m, c_n) \geq 0, \tag{12.33}$$

we have the following equations for probabilities:

$$P(A_1 = a_i, \ A_2 = a_j) = \sum_{klmn} f(a_i b_k c_l; a_j b_m c_n) \tag{12.34}$$

$$P(A_1 = a_i, \ B_2 = b_j) = \sum_{klmn} f(a_i b_k c_l; a_m b_j c_n) \tag{12.35}$$

$$P(C_1 = c_i, \ C_2 = c_j) = \sum_{klmn} f(a_k b_l c_i; a_m b_n c_j) \tag{12.36}$$

and analogous equations for $P(B_1 = b_i, A_2 = a_j)$, $P(A_2 = a_j, C_2 = c_j)$, $P(C_1 = c_i, B_2 = b_j)$, $P(C_1 = c_i, A_2 = a_j)$.

The number of unknown functions (12.33) is $2^6 = 64$, and there are $9 \times 4 = 36$ equations (Eqs. 12.34-12.36), and for eq(12.33) all f's are non-negative. Let us show that there exists a $|\psi\rangle$, for which these f do not exist!

Take the singlet state of the composite system of two spin 1/2 particles. We can write,

$$|\psi\rangle = 2^{-1/2}(|a_+\rangle_1 \otimes |a_-\rangle_2 - |a_-\rangle_1 \otimes |a_+\rangle_2) \tag{12.37}$$
$$|\psi\rangle = 2^{-1/2}(|b_+\rangle_1 \otimes |b_-\rangle_2 - |b_-\rangle_1 \otimes |b_+\rangle_2) \tag{12.38}$$
$$|\psi\rangle = 2^{-1/2}(|c_+\rangle_1 \otimes |c_-\rangle_2 - |c_-\rangle_1 \otimes |c_+\rangle_2). \tag{12.39}$$

STATISTICAL INTERPRETATION

Then, we have the following probabilities,

$$P(A_1 = a_i, A_2 = a_j) = 0 \tag{12.40}$$
$$P(B_1 = b_i, B_2 = b_j) = 0 \tag{12.41}$$
$$P(C_1 = c_i, C_2 = c_j) = 0. \tag{12.42}$$

Eq. (12.41), leads to the result, that the 16 variables on the right side of Eq. (12.34) being non-negative all become zero when $i = j$. Analogously we obtain from Eqs. (12.35)-(12.36) that 32 among the 64 functions f are zero, etc.

Only 8 variables are nonzero when all values are different for the subsystems.

$$f(a_+, b_+, c_+; a_-, b_-, c_-) = f_1 \tag{12.43}$$
$$f(a_+, b_+, c_-; a_-, b_-, c_+) = f_2 \tag{12.44}$$
$$f(a_+, b_-, c_+; a_-, b_+, c_-) = f_3 \tag{12.45}$$
$$f(a_+, b_-, c_-; a_-, b_+, c_+) = f_4 \tag{12.46}$$
$$f(a_-, b_+, c_+; a_+, b_-, c_-) = f_5 \tag{12.47}$$
$$f(a_-, b_+, c_-; a_+, b_-, c_+) = f_6 \tag{12.48}$$
$$f(a_-, b_-, c_+; a_+, b_+, c_-) = f_7 \tag{12.49}$$
$$f(a_-, b_-, c_-; a_+, b_+, c_+) = f_8 \tag{12.50}$$

Can these 8 functions satisfy Eqs. (12.34)-(12.35)? It is easy to see that the answer, in general, is no!

Really, writing the evident inequalities

$$f_3 + f_6 \geq 0 \tag{12.51}$$

$$(f_3 + f_4) + (f_2 + f_6) \geq f_2 + f_4 \tag{12.52}$$

we have, (taking zeros into account)

$$f_3 + f_4 = P(A_1 = a_+, B_2 = b_+) \tag{12.53}$$
$$f_2 + f_6 = P(B_1 = b_+, C_2 = c_+) \tag{12.54}$$
$$f_2 + f_4 = P(A_1 = a_+, C_2 = c_+). \tag{12.55}$$

But Eq. (12.52) leads to Bell's inequalities in d'Espagnat form.

$$P(A_1 = a_+, B_2 = b_+) + P(B_1 = b_+, C_2 = c_+) \geq P(A_1 = a_+, C_2 = c_+). \tag{12.56}$$

Let us calculate these probabilities of our system in the state $|\psi\rangle$ given by Eqs. (12.37)-(12.39). First, we write

$$|\psi\rangle = 2^{-1/2}(\sin\phi|a_+\rangle_1 \otimes |b_+\rangle_2 + \cos\phi|a_+\rangle_1 \otimes |b_-\rangle_2 \\ - \cos\phi|a_-\rangle_1 \otimes |b_+\rangle_2 + \sin\phi|a_-\rangle_1 \otimes |b_-\rangle_2). \qquad (12.57)$$

Then

$$P(A_1 = a_+, B_2 = b_+) = \frac{1}{2}\sin^2\phi. \qquad (12.58)$$

Then writing $|\psi\rangle$ as

$$|\psi\rangle = 2^{-1/2}(\sin\phi|b_+\rangle_1 \otimes |c_+\rangle_2 + \cos\phi|b_+\rangle_1 \otimes |c_-\rangle_2 \\ - \cos\phi|b_-\rangle_1 \otimes |c_+\rangle_2 + \sin\phi|b_-\rangle_1 \otimes |c_-\rangle_2) \qquad (12.59)$$

we obtain

$$P(B_1 = b_+, C_2 = c_+) = \frac{1}{2}\sin^2\phi. \qquad (12.60)$$

At last, writing

$$|\psi\rangle = 2^{-1/2}(\sin 2\phi|a_+\rangle_1 \otimes |c_+\rangle_2 + \cos 2\phi|a_+\rangle_1 \otimes |c_-\rangle_2 \\ - \cos 2\phi|a_-\rangle_1 \otimes |c_+\rangle_2 + \sin 2\phi|a_-\rangle_1 \otimes |c_-\rangle_2) \qquad (12.61)$$

we get

$$P(A_1 = a_+, C_2 = c_+) = \frac{1}{2}\sin^2 2\phi. \qquad (12.62)$$

So, from Eqs. (12.53)-(12.55) we must have

$$\frac{1}{2}\sin^2\phi + \frac{1}{2}\sin^2\phi \geq \frac{1}{2}\sin^2 2\phi, \qquad (12.63)$$

or

$$\cos^2\phi \leq \frac{1}{2} \qquad (12.64)$$

for $0 < \phi \leq \frac{\pi}{4}$ which, generally, is incorrect!

So, the adherents of the ensemble interpretation in this form must claim that:

a) some f can be negative;

b) that $A_1 = a_+$ if A is measured simultaneously with A_2 or C_2, but $A_1 = a_-$ if A is measured simultaneously with B_2. But this is just a *nonlocal* influence of a measurement of B_2 on A!

Notes

1. A detailed history of the *SIQT* can be found in Jammer[1]. See also Ballentine's article[2].
2. The validity of this statement depends on the validity of the ergodic hypothesis.
3. Concerning this controversy see [3,4].

References

[1] M. Jammer, *The Philosophical Foundations of Quantum Theory* (J. Wiley & Sons, New York, 1974).

[2] L. E. Ballentine, The statistical interpretation of quantum mechanics, *Rev. Mod. Phys.* **42**(4), 358–381 (1970).

[3] A. D. Alexandrov, On the meaning of the wave function (in Russian), *Doklady Akademii Nauk SSSR* **85**, 292–294 (1952)

[4] G. J. Mjakishev, What is the cause of the statistical character of quantum mechanics (in Russian), *Voprossy Filosofii* **6**, 146–159 (1954).

[5] C. Monroe et al, A "Schrödinger cat" superposition state of an atom, *Science* **272**, 1131–1136 (1996).

[6] M. Brune, E. Hagley, J. Dreyer, X. Maître, A. Maali, C. Wunderlich, S. Haroche and J. P. Raymond, Observing the progressive decoherence of the "meter" in a quantum measurement, *Phys. Rev. Lett.* **77**(24), 4887–4890 (1996).

[7] D. T. Gillespie, Untenability of simple ensemble interpretations of quantum measurements probabilities, *Am. J. Phys.* **54**(10), 889–894 (1986)

[8] M. Bitbol, An analysis of the Einstein-Podolski-Rosen correlations in terms of events, *Phys. Lett. A* **96**(2), 66–70 (1983).

13
NON-UNITARY EVOLUTION?

Different from *HVT*, there are some attempts to change the formalism of quantum theory, postulating new laws of evolution in time, i.e., by introducing a breakdown of the Schrödinger equation. In this case, the wave packet collapse arises as some natural process, described by this generalized equation, in such a way that contrary to quantum theory where there are two different processes of evolution, respectively by the Schrödinger equation and by measurement, here there is only one evolution process.

These attempts include:

(i) breakdown of the linearity of Schrödinger equation, putting some nonlinear term in it so that the superposition principle is broken. One of the first attempts in this direction is due to Wigner[1].

(ii) breakdown of the property of the Hamiltonian as Hermitian, in such a way that a non-Hermitian, dissipating part is added by hand and the unitarity of evolution is lost [2-7].

(iii) another interesting program consists of the replacement of the time-dependent Schrödinger equation by a stochastic equation (see [8] for a review). Recently a series of papers has appeared which relate stochasticity with the geometry of spacetime, particularly with torsion. The interested reader may consult [9] for details and also for an extensive list of references.

(iv) attempts by Gell-Mann, Hartle [10], Omnès [11] and others that instead of looking for a *generalized equation* of evolution, look at *history* as an *object* and then develop a generalization of quantum theory which gives *probabilities* for different histories (we discuss the Histories Approach in Chapter 14).

13.1 PRIGOGINE AND GHIRARDI ET AL.THEORIES

Here we mainly discuss the attempts (ii) attributed to Prigogine [7] and his school and to Ghirardi et al.[2-6] and which have been investigated for a long period.

According to Prigogine ideas, at a fundamental level in nature, there is an irreversibility of time. There is then, a strong reason to change the Schrödinger equation, which, as is well known, is time reversible.

The new term, added to the Schrödinger equation, must be small enough in order not to contradict the experimental evidences supporting standard quantum physics, but it must play an important role concerning macroscopic systems, containing an enormous (Avogadro) number of quantum particles.[1] It is this term, that, e.g., prohibits the existence of the superposition state of the dead and alive Schrödinger cat for a more or less a long time. This difference in description of micro and macro systems is in some sense close to Zurek [13] and Omnès [11] *environment* idea, which we discussed previously in Chapter 2. We recall that according to this idea, there are *collective variables* for macrosystems which behave in such a manner, that due to interaction with the environment, the density matrix for them on some special basis, becomes rapidly diagonalized and interference terms are not observed.

But, different from this idea, the authors of the changing Schrödinger equation agree[2-7] with Bell's criticism of an *environmental* solution of the Schrödinger cat paradox. Bell said: "Here are some words, which however legitimate and necessary in applications, have no place in a formulation with any pretension to physical precision: system, apparatus, environment ... The concepts *system, apparatus, environment* immediately imply an artificial division of the world" (see [14]). And further, "An appeal has been made that goes beyond the ordinary Schrödinger equation, to an a priori split of a physical system into a macroscopic system, detector and environment. But no rules have ever been given for making such a split, and certainly a physical system does not come with a subsystem containing a little sign reading, "I am the environment: Trace over me. Without such rules one cannot in the general case apply the environment trace prescription to determine what is desired, namely the *preferable basis* states that one can actually observe. Thus, the whole scheme appears devoid of fundamental significance."

The most developed scheme for breaking the unitarity of the evolution operator can be found in Ghirardi, Rimini, Weber (*GRW*) papers [2-6] who developed the so called *continuous spontaneous localization theory*. A new term is added to the Schrödinger equation which depends upon a *randomly fluctuating field*, distinguishing states that *differ* from one another in particle number density anywhere in space. If the differences in particle number density are large enough the superposition rapidly evolves to one or another of these states (depending upon what fluctuation actually occurs). The dynamics thereby determines the observed macroscopic preferred basis states in a well defined way. If there are small differences in particle number density, as in the case of a microscopic system, the reduction takes place at a *negligible rate*.

Instead of the Schrödinger equation (or equivalently, the equation for the density matrix), which contains only the Hamiltonian operator, *GRW* propose to write the following generalized equation

$$\frac{d\rho}{dt} = -\frac{i}{\hbar}[\mathbf{H}, \rho] + \lambda \sum_k (\mathbf{T}_k(\rho) - \rho) \qquad (13.1)$$

where ρ is the density matrix for the n-particles system and

$$\mathbf{T}_k[\rho] = \int d\vec{x}\, \mathbf{L}_{\vec{x}}^k \rho \mathbf{L}_{\vec{x}}^k, \qquad (13.2)$$

and there is for each particle "k" a localization process

$$|\psi\rangle \to |\psi_{\vec{x}}^k\rangle = \mathbf{L}_{\vec{x}}^k |\psi\rangle \qquad (13.3)$$

where $|\psi\rangle$ is the state vector of the n-particles system and $\mathbf{L}_{\vec{x}}^k$ is the operator

$$\mathbf{L}_{\vec{x}}^k = \left(\frac{\alpha}{r}\right)^{3/4} \exp\left[-\left(\frac{\alpha}{2}\right)(\mathbf{q}_k - \vec{x})^2\right] \qquad (13.4)$$

where $\mathbf{q}_k$ is the position operator of the k-th particle.

Then, new fundamental constants must exist in nature which are proposed to be

$$\lambda = 10^{-6} s^{-1}, \quad \frac{1}{\sqrt{\alpha}} = 10^{-5} cm. \qquad (13.5)$$

Looking at Eq. (13.1) it is easy to see that:

(a) It breaks unitarity of the evolution, because λ is real, and is then similar to a dissipation term. However, different from standard reasoning, it is present not because of our *ignorance* concerning the environment, but at the fundamental level, leading to a *non conservation* of the energy.

(b) There is some *stochasticity* as the random field in standard probability theory, leading to the breakdown of the *entanglement* for superpositions in such a way that *mixtures* instead of superpositions can arise.

We can also say that property (a) is enough to achieve the goal: transform superpositions into mixture.

For an n-particle system, *GRW* propose yet another form of eq. (13.1):

$$\frac{d\rho}{dt} = -\frac{i}{\hbar}[\mathbf{H}, \rho] - \frac{\gamma}{2} \int d\vec{x}\, [\mathbf{N}(\vec{x}), [\mathbf{N}(\vec{x}), \rho]] \qquad (13.6)$$

where $\lambda = \gamma \left(\frac{\alpha}{4r}\right)^{\frac{3}{2}}$, and $\mathbf{N}(\vec{x})$ is the number density operator for a system of identical particles

$$\mathbf{N}(\vec{x}) = \left(\frac{\alpha}{2r}\right)^{\frac{3}{2}} \sum_s \int d\vec{y}\, \exp(-\frac{\alpha}{2}(\vec{y}-\vec{x})^2) \mathbf{a}^\dagger(\vec{y}, s) \mathbf{a}(\vec{y}, s) \qquad (13.7)$$

where $\mathbf{a}^\dagger(\vec{y}, s)$ and $\mathbf{a}(\vec{y}, s)$ are Fock's creation and annihilation operators at point $\vec{y}$ and s is the spin.

From Eq. (13.6) we can see that a new term, depending on γ now, breaks *Pauli's principle* for identical particles, giving *names* to them. The new term also depends on the particle density $\mathbf{N}(\vec{x})$ and is different at different points. Then, it is easy to understand that in this theory, superpositions of states of a macroscopically great number of particles will become mixtures.

In standard quantum theory this is not the case: we can locate one particle in an entangled n-particle state, but this will not lead to the disentanglement (or *dragging* effect) for all the other $(n-1)$ particles. They still will be in superpositions.

Going from the standard coordinate operators $\mathbf{q}_i$ to the centers of mass-coordinate operator $\mathbf{q}$ and the relative coordinate operators $\mathbf{r}_i$

$$\mathbf{q}_i = \mathbf{q} + \sum_{j=1}^{n-1} c_{ij} \mathbf{r}_j \tag{13.8}$$

GRW show, that if we are only interested in the center of mass motion, then, as in the *environmental approach,* by tracing over the relative coordinate degrees of freedom, we obtain for the density matrix $\rho_\mathbf{q}$ from Eq. (3.1), the equation

$$\frac{d\rho_q}{dt} = -\frac{i}{\hbar}[\mathbf{H}_q, \rho_q] + \lambda \mathbf{N}(\mathbf{T}_q - \rho_q) \tag{13.9}$$

Then, instead of a small constant λ, a new constant $\lambda_{\text{macro}} = \lambda n$ with n of the order of Avogadro's number, arises. In this way, we can understand why for macroscopic systems with n-large enough the effect is present, while for one microparticle it is small. Károlyhazy, Frenkel and Lukás[15], Maximov[16], Diosi[17] and also Penrose[18] proposed the idea that instead of introducing new fundamental constants we might think of using as constants playing the role of λ, e.g., the gravitational constant or Planckean mass $m_p = 10^{-5}g$. Maximov[16] developed the idea of Infeld[19], that the gravitational field is not a physical field, but is pure geometry and so it does not need to be quantized. Nevertheless, in this approach, energy conservation for standard particles does not hold when they are in the interaction with geometrical gravity, which then is not characterized by any energy at all.

Anyway, in this approach the unitarity of the quantum theory can be broken down by gravity, just because of a non-quantizing gravity, which is always described by a c-number field and not by an operator field.

Contrary to this idea, Penrose[18] suggests the idea (although superficially developed) that if we try to quantize spacetime, there is a problem of interpretation about the meaning of the concept of *superposition of spacetimes*. As a result, *timelike loops,* or *time machines* can arise, implying a breakdown of unitarity in quantum theory.

The idea that quantum gravity can help to solve the Schrödinger's cat paradox due to the possible existence of *wormholes* (a similar idea can also be found in Hawking [20]) which lead to the appearance of some random field, like in *GRW* theory, was also discussed by Ellis and Nanoupolos[21]. Again, multiplication of a small constant with a large number N_A for the center of mass motion was exploited.

13.2 CRITICISMS

The general critical remarks to *GRW* theory (and similar theories) are the following:

(a) It not only lacks energy conservation, but is nonlocal like any *physical* theory of the wave packet collapse. So, there are serious problems with its relativistic generalization which are of the same nature as in de Broglie-Bohm hidden variables theory.

(b) Nonlocality leads to serious divergence problems for calculation in relativistically generalized versions of *GRW* theory.

(c) Attempts to use gravitation as a *random field*, transforming superpositions into mixtures if based on equations like the ones of *GRW* theory, contradict the *equivalence principle* which says that the quantum theory of particles, interacting with gravitation, is to be constructed as a quantum theory in curved space-time. But, in this theory, the only thing that is done is to change the metric in Klein-Gordon-Fock or Dirac equations [22] without introducing dissipative terms. Dissipative terms can arise only if process of particle-antiparticle creation or the *eating* of particles by the gravitational field arises [22]. But calculations[22] show that such terms must be very small for standard particles and it is difficult to construct the very ambitious *GRW* claim on the basis of such effects.

(d) There is the experimental possibility of checking all these theories, that claim that interference disappears for a large number of particles in the so-called *macroscopic quantum mechanics* of systems, such as superconducting quantum interference devices (*SQUIDs*), squeezed light, etc., when a great number of particles are in the entangled state (*Schrödinger's cat* states in quantum optics [23]). But no experimental evidence contradicting the predictions of standard quantum theory has been observed until now.

Nevertheless, the possibility of experimentally falsifying these theories is one of their strong features, and it may happen that future experiments on the *frontier* between quantum and classical physics will show something like the *GRW* effect.

We end this chapter by quoting the new Prigogine's book[24]. There, Prigogine defends vigorously a realistic interpretation for quantum mechanics. He insists that this can be achieved only by a *paradigmatic* change. This implies in the recognition that the irreversibility of time is a fundamental aspect of the world we live in and then it is necessary since the beginning to formulate physical laws that are not time reversal invariant. Such an approach cannot be pursued with the standard mathematical methods of quantum mechanics. In particular it is necessary to use functional spaces more general then Hilbert spaces. In his theory a statistical description (in the sense defined in Chapter 12) is necessary, but the criticisms presented there do not apply to Prigogine's theory, where the statistical description is a priori supposed the fundamental description of reality and this has nothing to do with our ignorance concerning phenomena. Indeed, the statistical interpretation itself in this approach suggest that we must recognize that in the usual quantum theory there exists non integrable systems due to the existence of Poincaré resonances.

Notes

1. Recently Gähler [12] found extremely low upper bounds for the magnitude of the nonlinear term in an important family of nonlinear Schrödinger equations.

References

[1] E. P. Wigner, in: *The Scientist Speculates,* edited by I. J. Good (Heinemann, London, 1961), pp. 284–302.

[2] G. C. Ghirardi, A. Rimini and T. Weber, Unified Dynamics for microscopic and macroscopic systems *Phys. Rev. D* **34**(2), 470–191 (1986).

[3] G. C. Ghirardi, R. Grassi and T. Weber, Continuous spontaneous reduction model involving gravity, *Phys. Rev. A* **42**(3), 1057–1065 (1990).

[4] G. C. Ghirardi, R. Grassi and P. Pearle, Relativistic dynamic reduction models—general framework and examples *Found. Phys.* **20**(11), 1271–1316 (1990).

[5] G. C. Ghirardi, R. Grassi and P. Pearle, Negotiating the tricky border between quantum and classical, *Phys. Today* **46**(4), 13–15 (1993).

[6] G. C. Ghirardi, Properties and events in a relativistic context:revisiting the dynamical reduction program, *Found. Phys. Lett.* **9**(4), 313–355 (1996).

[7] I. Prigogine, *Non Equilibrium Statistical Mechanics* (Interscience Publishers, New York, 1962).

[8] M. Jammer, *The Philosophy of Quantum Mechanics* (J. Wiley & Sons, New York, 1974).

[9] D. L. Rapoport, Riemann-Cartan-Weyl quantum geometries.1. Laplacians and supersymmetrical systems, *Int. J. Theor. Phys.* **35**(2), 287–309 (1996).

[10] M. Gell-Mann and J. B. Hartle, in: *Proceedings of the 3rd International Symposium: Foundations of Quantum Mechanics in the Light of New Technology,* edited by S. Kobayashi, H. Ezawa, Y. Murayama and S. Nomura (Physical Society of Japan, Tokyo, 1990), pp..

[11] R. Omnès, Consistent interpretations of quantum mechanics, *Rev. Mod. Phys.* **64**(2), 339–382 (1992).

[12] J. Felber, R. Gähler and R. Golub, Test of the time dependent Schrödinger equations with very slow neutrons, *Physica B&C* **151**(1-2),135–139 (1988).

[13] W. H. Zurek, Decoherence and the transition from quantum to classical, *Phys. Today* **44**(10), 36–44 (1991).

[14] J. S. Bell, in: *Sixty-two years of Uncertainty: Historical, Philosophical and Physical Inquires in the Foundations of Quantum Mechanics,* edited by A. Miller (Plenum Press, New York, 1990), pp.17–31.

[15] A. Károlyházy, A. Frenkel and B. Lukács, in: *Quantum Concepts in Space and Time,* edited by R. Penrose and C. J. Isham (Clarendon Press, Oxford,1986), pp.109–128.

[16] Yu. M. Maximov, On objective interpretation of the wave function in quantum mechanics, *Izvestija Vusov SSSR, Fizika* **5**, 63–69 (1976).

[17] L. Diosi, Models for universal reduction of macroscopic quantum fluctuations, *Phys.Rev. A* **40**(3), 1165–1174 (1989).

[18] R. Penrose, *Shadows of the Mind* (Oxford University Press, Oxford, 1994).

[19] L. Infeld, *Fortschr. Phys.*, Über die jüngste entwicklung der klassischen elektrodynamik, **1**(2) 88–98 (1953).

[20] S. W. Hawking, The unpredictability of quantum gravity, *Commun. Math. Phys.* **87**(3), 395–415 (1982).

[21] J. Ellis, S. Mohanty and D. Nanoupolos, Quantum gravity and the collapse of the wave function, *Phys. Lett. B* **221**(2), 113–119 (1989).

[22] A. A. Grib, S. G. Mamayev and V. M. Mostepanenko, *Vacuum Quantum Effects in Strong Fields* (Friedmann Laboratory Publishing, St. Petersburg, 1994).

[23] A. J. Leggett, Scrödinger cat and its laboratory cousins, *Contemp. Phys.* **25**(6), 583–598 (1984).

[24] I. Prigogine, *La Fin des Certitudes. Temp, Chaos et les Lois de la Nature* (Éditions Odile Jacob, Paris, 1996).

14
HISTORIES APPROACH

14.1 WHAT IS A HISTORY?

The idea of uniting the two types of wave function change: one based on the Schrödinger equation and the other on the wave packet collapse and of forming one "objective" evolution, inspired Gell-Mann and Hartle[1] and Omnès[2] to propose the so called *histories approach*.

In a sense, these authors try to unite some ideas of the *many worlds interpretation*, Stapp's[3] *objectively occurring* collapses and Feynman and Vernon[4] and Zurek [5] *environmental decoherence* interpretations.

Let **H** be the Hamiltonian of the system. The unitary Schrödinger evolution operator is

$$\mathbf{U}(t) = \exp[-i\frac{\mathbf{H}t}{\hbar}]. \qquad (14.1)$$

In order to introduce the concept of history, it is better to use the Heisenberg representation, where the density matrix of the system will be time independent and the observables will be time dependent. If to the observable A corresponds the Hermitian operator **A** in the Schrödinger representation, then in the Heisenberg representation, we have the time dependent operator

$$\mathbf{A}(t) = \mathbf{U}^{-1}(t)\mathbf{A}\mathbf{U}(t) \qquad (14.2)$$

The main object of the approach is *history*, which is a time ordered ($t_1 < t_2 < ... < t_n$) *conjunction* of properties defined by the observables $\{A_1, A_2, ... A_n\}$. The properties do not need to be compatible. It is to histories that probabilities are assigned. To see how this is done, we first assume that:

(i) the initial state of the quantum system at time $t_0 (< t_1)$ is given by the density matrix ρ (in the Heisenberg representation) and

(ii) the spectrum of each observable A_i represented by the operators $\mathbf{A}_i(t_i)$ is divided into a complete family of disjoint sets $D_i^{\alpha_i}$.

Let $\mathbf{P}_i^{\alpha_i}(t_i), i = 1, ..., n$ be the time dependent projectors on the subspace spanned by the eigenvectors of the operators $\mathbf{A}_i$ with eigenvalues in the range $D_i^{\alpha_i}$.

Given the set $\{\alpha_i\}$, we define the history[1] $\mathfrak{A}$ by the time ordered sequence of properties,

$$\mathfrak{A} = \{\mathbf{P}_1^{\alpha_1}(t_1), \mathbf{P}_2^{\alpha_2}(t_2), ... \mathbf{P}_n^{\alpha_n}(t_n)\} \quad (14.3)$$
$$t_1 < t_2 < ... < t_n.$$

The joint probability of finding all the properties in an appropriate sequence of measurements is called the probability of the history and is given by[2]

$$p_{\mathfrak{A}} = Tr(\mathbf{P}_n^{\alpha_n}(t_n)...\mathbf{P}_2^{\alpha_2}(t_2)\mathbf{P}_1^{\alpha_1}(t_1)\rho \mathbf{P}_1^{\alpha_1}(t_1)...\mathbf{P}_n^{\alpha_n}(t_n)). \quad (14.4)$$

This equation is the well known Wigner's formula for the probabilities[7]. By varying the set $\{\alpha_i\}$ we obtain a *complete family* of histories.

14.2 CONSISTENT HISTORIES

The probabilities for histories are additive for *disjoint* properties occuring at the same time. The probabilities of a *larger* history is the sum of the probabilities for the more detailed ones entering it.

However, additivity is not satisfied by all complete families of histories, since the probabilities for histories must be consistent with the quantum additivity of amplitudes. This condition is expressed by the so-called *consistency conditions*, first found by Griffiths in 1984[8]. Gell-Mann and Hartle presented the consistency conditions as

$$Tr\{\mathbf{P}_{n-1}^{\alpha_{n-1}}(t_{n-1})...\mathbf{P}_1^{\alpha_1}(t_1)\rho \mathbf{P}_1^{\alpha_1'}(t_1)...\mathbf{P}_{n-1}^{\alpha_{n-1}'}(t_n)\} = 0 \quad (14.5)$$

where the sequence $\{\alpha_i\}$ is different from the sequence $\{\alpha_i'\}$ ($i = 1, ..., n - 1$), but it turns out that these conditions are sufficient for additivity but they are not necessary. The necessary conditions have been given by Griffiths [8] and Omnès[2]. For the simplest case $n = 2$, Griffiths proposed :

$$ReTr\{\mathbf{P}_1^{\alpha_1}(t_1)\rho \mathbf{P}_1^{\alpha_1'}(t_1)\mathbf{P}_2^{\alpha_2}(t_2)\} = 0 \quad (14.6)$$

and Omnès proposed

$$Tr([\mathbf{P}_1^{\alpha_1}(t_1), [\rho, \mathbf{P}_1^{\alpha'_1}(t_1)]]\mathbf{P}_2^{\alpha_2}(t_2)) = 0 \qquad (14.7)$$

In the case of a spin 1/2 system, if the initial state has a definite value of the spin component along the direction $\vec{n}_0$ and the properties at times t_1 and t_2, correspond respectively to the values of the spin components along two direction $\vec{n}_1, \vec{n}_2$, then Eqs. (14.6)-(14.7) give

$$(\vec{n}_0 \times \vec{n}_1) \cdot (\vec{n}_1 \times \vec{n}_2) = 0 \qquad (14.8)$$

while Eq. (14.5) gives

$$\vec{n}_1 = \pm \vec{n}_0, \text{ or } \vec{n}_1 = \pm \vec{n}_2. \qquad (14.9)$$

Histories satisfying the consistency conditions are said to be *consistent histories*. All other histories are said to be inconsistent.

To consistent histories we can give *yes-no* values and say that they *are*! To inconsistent histories we cannot give *yes-no* values.

If properties are compatible, then they are consistent with respect to every initial state.

An important feature of the approach is that for special states there may be properties that are consistent but not compatible. An example of consistency is when the state is a probabilistic mixture of pure states $|\phi_a\rangle$ with weights w_a and the projectors $\mathbf{P}_i$ project on $|\phi_i\rangle$ while $\mathbf{Q}_j$ are arbitrary projectors. The probabilities of the conjunctions, taken in order are

$$p(\mathbf{P}_i, \mathbf{Q}_j, \mathbf{P}_i) = w_i \langle \phi_i | \mathbf{Q}_j | \phi_i \rangle. \qquad (14.10)$$

At this point, we can introduce the main claim of the supporters of the histories approach:

Consistent histories replace the reduction of state vectors.

Let us analyze this claim. Recall that if our quantum system is in the state $|\Psi\rangle$, then the probability of $\mathbf{P}_i$ and then $\mathbf{Q}_j$ is

$$p(\mathbf{P}_i, \mathbf{Q}_j, \mathbf{P}_i) = \langle \Psi | \mathbf{P}_i \mathbf{Q}_j \mathbf{P}_i | \Psi \rangle = \| \mathbf{Q}_j \mathbf{P}_i | \Psi \rangle \|^2 . \qquad (14.11)$$

The conditional probability of the $\mathbf{Q}_j$ given $\mathbf{P}_i$ is the quotient

$$p(\mathbf{P}_i, \mathbf{Q}_j, \mathbf{P}_i)/p(\mathbf{P}_i) = \| \mathbf{Q}_j \mathbf{P}_i | \Psi \rangle \|^2 / \| \mathbf{P}_i | \Psi \rangle \|^2 . \qquad (14.12)$$

This is the probability of $\mathbf{Q}_j$, that would have been given by *reduction* of the state vector if the previous measurement had made $\mathbf{P}_i$ true.

Omnès, Gell-Mann and Hartle tried to show that classical properties, defined by collective variables define consistent (or *decohering*) histories.

Omnès calls the classical properties of macroscopic bodies arising from his theory, *phenomena*. Classical properties are called *facts* if they actually occur in nature. The passage from *phenomena* to *facts* is called *actualization*. The theory has *no* explanation at all for *actualization,* but this does not worry Omnès, who said:"The existence of actual facts can be added to the theory from outside as a supplementary condition issued from empirical observation."

At first it seems as if only *classical properties* really exist as in the *absence of quantum objects* interpretation of quantum theory, but according to the theory of measurement, it becomes possible to include some properties of microobjects in the consistent histories.

This is done as follows. Let us have a measuring apparatus and let us call *data* its classical properties. The apparatus interacts with some atomic system whose properties are called *results*. The *results* precede the *data*. In appropriate circumstances, *results* and *data* fit into consistent histories and this constitutes a *measurement*. A *factual* property is always regarded as being true and according to Omnès, if in the context of a measurement the property expressing the datum is a fact, then the property expressing the result must also be considered as true. In this way, we are allowed in special circumstances (and only in these), to speak about the *truth* of the properties of systems on an atomic level.

The Hilbert space of the apparatus and atomic system is $\mathcal{H} = \mathcal{H}_A \otimes \mathcal{H}_S$.

Let the apparatus, which in the present case is the collective system, be initially in some state $|\phi_0\rangle \in \mathcal{H}_A$ and the atomic system in the superposition $\sum c_i |\chi_i\rangle \in \mathcal{H}_S$. Then the combined system $(\mathcal{A} + \mathcal{S})$ is in the state

$$|\Psi\rangle = |\phi_0\rangle \otimes \sum c_i |\chi_i\rangle. \qquad (14.13)$$

A measurement is a special interaction, described by the Schrödinger evolution, so that always

$$\mathbf{U}(|\phi_0\rangle \otimes |\chi_i\rangle) = |\phi_i\rangle \otimes |\psi_i\rangle. \qquad (14.14)$$

Then the resulting state vector for the combined system is

$$\mathbf{U}|\Psi\rangle = \sum_i c_i |\phi_i\rangle \otimes |\psi_i\rangle. \qquad (14.15)$$

Let $\mathbf{P}_i$ and $\mathbf{Q}_j$ be projectors, corresponding to the *results* and *data* at time zero. As is well known in the Heisenberg picture the state vector remains the initial $|\psi\rangle$ while the projectors for the *data* at the later time are $\mathbf{Q}_j(t) = \mathbf{U}^{-1} \mathbf{Q}_j \mathbf{U}$. Using the fact that $\mathbf{P}_i$ projects on $|\chi_i\rangle$ we obtain for a *history*:

$$\begin{aligned} p(\mathbf{P}_i, \mathbf{Q}_j(t), \mathbf{P}_i) &= c_i^* c_k \langle \phi_0 | \otimes \langle \chi_i | \mathbf{Q}_j(t) |\phi_0\rangle \otimes |\psi_k\rangle \\ &= c_i^* c_k \langle \phi_i | \otimes \langle \psi_i | \mathbf{Q}_j |\phi_k\rangle \otimes |\psi_k\rangle. \end{aligned} \qquad (14.16)$$

However, $\mathbf{P}_j$ projects on $|\phi_j\rangle$ and since the $|\phi_j\rangle$ are orthogonal, we get

$$p(\mathbf{P}_i, \mathbf{Q}_j(t), \mathbf{P}_i) = |c_i|^2 \delta_{ij}\delta_{jk}. \tag{14.17}$$

This probability is zero for $i \neq k$ and then, the probability for the ordered conjunction of *result* and *datum* is

$$p(\mathbf{P}_i, \mathbf{Q}_j(t), \mathbf{P}_i) = |c_i|^2 \delta_{ij}. \tag{14.18}$$

This means that each *result* is equivalent to the corresponding *datum* and has the same probability $p(\mathbf{P}_i) = p(\mathbf{Q}_j(t)) = |c_i|^2$.

In this way, like in the Copenhagen interpretation, it is because of the *interaction* with the apparatus (described by quasi-classical properties) during a measurement, that microproperties acquire *yes-no* meaning.

Nevertheless, here the observer and his consciousness does not play *any* important role.

The *EPR* situation for two particles is treated without using such words as *nonlocality* or *nonseparability*. Again, as in the *absence of existence of quantum objects* interpretation, it is just said that there exists a consistent history with *unexplained and unexplainable coincidences* for random events at space-like distances, manifested, e.g., in Aspect et al experiments.

14.3 CRITICISMS

1. The notion of probability for a family of histories is far from any experimental foundation contrary to the usual frequencies approach, etc.

2. It was shown by Dowker and Kent [9] (see also Diosi [10]) that for finite dimensional Hilbert space, there exists a finite number $\leq r_n$ (where n is the dimension of the Hilbert space), so that the number of freely chosen $\alpha_1...\alpha_n$ in the sequence of properties cannot be larger than this finite number. This means that, somehow, randomness in quantum mechanics and indeterminism (or free will) totally disappear after some finite sequence of events. Surely this strange property totally contradicts what we have in ordinary quantum mechanics. The only escape is to break the *consistency* condition and to go to *inconsistent* histories, but then the whole scheme must be changed.

3. Truth values in the histories approach are given not to *events*, but to histories as sequences of events. This can lead to problems concerning truth values for events. Really, it is possible that up to some moment t_1, we have some sequence of events which can be prolonged in the future in many different ways, corresponding to different quasi-classical consistent histories. But then, if one history is considered as *true* and the other alternative as *false*, what can we say about the truth of an event belonging to both? Is it both *true* and *false*?

The solution would be the uniqueness of the prolongation into the future, but it was shown in [11] that there exist many different prolongations. So, we must have some additional principle (like final density matrix) to maintain this uniqueness, which then leads to absence of any free will and choice and seems very distant from our intuitive notions.

4. *Problem of the uniqueness of the classical world.* Gell-Mann and Hartle tried to prove the uniqueness of the classical evolution of the classical world ("the Moon always follows its orbit"). In order to do this, it was necessary to go to cosmology and speculate about the initial density matrix (or the wave function) of the universe as we have said previously. In order to have a *classical* world, Gell-Mann and Hartle use *coarse-grained projection operators*. To have a real classical history, they assume an objective limit for the degree of fine grainedness of these operators (our old Schrödinger's cat paradox!). Also, there is a possibility, like in Everett's *many worlds* picture, of *many-pasts,* since there may exist many consistent histories, leading to what we see at the present.

All these problems are far from being solved today. Nevertheless, the *histories approach* as well as the *many worlds* interpretation seems very attractive to cosmologists, speculating on the possibility of using quantum theory for the whole universe.

We think it is worth ending this Chapter with some observations from Farris' review [11] of Omnès version of the histories approach presented in his new book "The Interpretation of Quantum Theory"[12]: "What can we conclude from such a book? Some physicists regard quantum mechanics as a totally successful theoretical framework; they consider any attempt to raise questions about its foundations[3] as irritating distraction. Omnès, to his credit, recognizes that the puzzles are profound and that the traditional resolutions do not achieve the level of clarity appropriate to a complete science. Furthermore, he embarks on the task of providing a resolution within the framework of orthodox quantum mechanics. If he or anyone else succeeds in this task, then all questions should be settled, once and for all. In the present case, the book is energetic and lively and full of examples and ideas. But the core resolution is merely a desperate bluff. The resolution is to "add to the whole logical construction an assumption according to which present phenomena are unique (and therefore facts)." The bluff lies in such statements as: "the actuality of facts is something that need not be explained by the theory," and "when one finds a gap between theory and reality only at their common extremities, this is not a failure but the mark of an unprecedented success for quantum mechanics, as compared with the theories before it." The fact that an obviously competent physicist is driven to such assertions is evidence that the quantum theory remains in conceptual murk. The challenge remains: interpret quantum mechanics on its own terms, without appeal to authority, in a way that makes sense to reasonable people. This challenge has not met an adequate response."

Notes

1. The idea of a history can be generalized from the discrete case to the continuous one (Isham and Linden [6]), but will not be discussed here.

2. See [2] for the justification of this formula.

3. This is an authors' note: in connection with this statement, see the recent debate between Hawking and Penrose [13].

References

[1] M. Gell-Mann and J. B. Hartle, Classical equations for quantum systems, *Phys. Rev. D* **47**(8), 3345–3382 (1993).

[2] R. Omnès,, Consistent interpretations of quantum mechanics, *Rev. Mod. Phys.* **64**(2), 339–382 (1992).

[3] H. P. Stapp, *Mind, Matter, and Quantum Mechanics* (Springer-Verlag, Berlin, 1993).

[4] R. P. Feynman and F. L. Vernon, Jr., The theory of a general quantum system interacting with a linear dissipative system, *Ann. Phys. (N. Y.)* **24**(1), 118–130 (1963).

[5] W. H. Zurek, Pointer basis of quantum apparatus—into what mixture does the wave packet collapse, *Phys. Rev. D* **24**(6), 1516–1525 (1981).

[6] C. J. Isham and N. Linden, Continuous histories and the history group in generalized quantum theory, *J. Math. Phys.* **36**(6), 5392–5408 (1995).

[7] B. d'Espagnat, *Veiled Reality. An Analysis of Present-Day Quantum Mechanical Concepts* (Addison-Wesley Publishing Co., Reading, MA, 1995).

[8] R. B. Griffiths, Consistent histories and the interpretation of quantum mechanics, *J. Stat. Phys.* **36**(1-2), 219–272 (1984).

[9] F. Dowker and A. Kent, Properties of consistent histories, *Phys. Rev. Lett.* **75**(17), 3038–3041 (1995).

[10] L. Diosi, Models for universal reduction of macroscopic quantum fluctuations, *Phys. Rev. A* **40**(3), 1165–1174 (1989).

[11] W. Farris, Review of Roland Omnès, "The interpretation of quantum mechanics", *Notices of the Am. Math. Soc.* **43**(11), 1328–1339 (1996).

[12] R. Omnès, *The Interpretation of Quantum Mechanics* (Princeton University Press, Princeton, NJ, 1994).

[13] S. W. Hawking and R. Penrose, *The Nature of Space and Time* (Princeton University Press, Princeton, NJ, 1996).

15
QUANTUM MIRACLES AND CRYPTOGRAPHY

Besides the breaking of some inequalities, nonlocality in quantum physics leads to some unusual correlations between separated events, which from an ordinary point of view, seem like *miracles*. This means that for someone, not knowing quantum theory and using standard logic and probability theory, these correlations seem totally improbable. We now study some of these *miracles*.

15.1 GREENBERGER ET AL. EXAMPLE

Greenberger, Horn, Shimony and Zeilinger[1] obtained for the entangled state vector of systems of 3, 4 or more particles, some relations (*GMSZ*-relations) which are valid in quantum theory and invalid in standard probability theory.

Suppose we have a system consisting of four particles with spin $1/2$: particles 1, 2 move, e.g., in the positive z-direction and particles 3, 4 in the negative z-direction. Suppose that these particles are the products of the decay of a spin 0 particle into two particles with spin 1, and that each of these spin 1 particles decayed into a pair of spin $1/2$ particles, with the z-component of the spin equal to zero. Then the state of these four particles is described by a vector $|\psi\rangle \in \mathcal{H}^1 \otimes \mathcal{H}^2 \otimes \mathcal{H}^3 \otimes \mathcal{H}^4$, given by

$$|\psi\rangle = \frac{1}{\sqrt{2}} \left(|+\rangle_1 \otimes |+\rangle_2 \otimes |-\rangle_3 \otimes |-\rangle_4 - |-\rangle_1 \otimes |-\rangle_2 \otimes |+\rangle_3 \otimes |+\rangle_4 \right), \quad (15.1)$$

and where each $\mathcal{H}^i$, $i = 1, 2, 3, 4$ is a copy of the two dimensional spin space, that we have already encountered many times in this book.

We suppose that these four particles go through four Stern-Gerlach apparatuses, in such a way that the spin projections on given axes $\vec{n}_i$ in the xy plane are measured. For each Stern-Gerlach apparatus some polarization angle ϕ_i is fixed. If $\mathbf{S}_{n_i}, i = 1, 2, 3, 4$ denotes the spin operators in the $\vec{n}_i$ directions, then the quantum mechanical expectation value is:

$$E^\psi(\vec{n}_1, \vec{n}_2, \vec{n}_3, \vec{n}_4) = \langle\psi| \otimes_{i=1}^{4} \mathbf{S}_{n_i} |\psi\rangle = -\cos(\phi_1 + \phi_2 - \phi_3 - \phi_4). \quad (15.2)$$

If

$$\phi_1 + \phi_2 - \phi_3 - \phi_4 = 0, \quad (15.3)$$

then

$$E^\psi(\vec{n}_1, \vec{n}_2, \vec{n}_3, \vec{n}_4) = -1. \quad (15.4)$$

If

$$\phi_1 + \phi_2 - \phi_3 - \phi_4 = \pi, \quad (15.5)$$

then

$$E^\psi(\vec{n}_1, \vec{n}_2, \vec{n}_3, \vec{n}_4) = +1. \quad (15.6)$$

From Eqs. (15.3)-(15.6), it is clear that there is a correlation such that the knowledge of the outcomes of the apparatuses 1, 2, 3 (+1 meaning up, −1 meaning down) for known angles, gives knowledge of the result that will be detected by apparatus 4. We can say that if we could have an analogous situation in human society, then, e.g., in a given election, if three persons vote *for* a given candidate, the fourth necessarily will vote *against* the candidate in the case given by Eq. (15.4) and *for* in the case given by Eq. (15.5).

This correlation, however, cannot be obtained from any classical law and is a kind of *quantum miracle*! Indeed, suppose, as in the proof of Bell's inequalities, the existence of functions $A_\lambda(\phi_1), B_\lambda(\phi_2), C_\lambda(\phi_3), D_\lambda(\phi_4)$ with values ±1 dependent on a hidden variable λ (defined by the general *past*).

If Eq. (15.3) is valid, then

$$A_\lambda(\phi_1)B_\lambda(\phi_2)C_\lambda(\phi_3)D_\lambda(\phi_4) = -1. \quad (15.7)$$

If eq.(15.5) is true, then

$$A_\lambda(\phi_1)B_\lambda(\phi_2)C_\lambda(\phi_3)D_\lambda(\phi_4) = 1. \quad (15.8)$$

Let us show that Eq. (15.7) and Eq. (15.8) are contradictory.
From Eq. (15.7), we have

$$\begin{align}
A_\lambda(0)B_\lambda(0)C_\lambda(0)D_\lambda(0) &= -1 & (15.9)\\
A_\lambda(\phi)B_\lambda(0)C_\lambda(\phi)D_\lambda(0) &= -1 & (15.10)\\
A_\lambda(\phi)B_\lambda(0)C_\lambda(0)D_\lambda(\phi) &= -1 & (15.11)\\
A_\lambda(2\phi)B_\lambda(0)C_\lambda(\phi)D_\lambda(\phi) &= -1. & (15.12)
\end{align}$$

From Eqs. (15.9)-(15.10) we have

$$A_\lambda(\phi)C_\lambda(\phi) = A_\lambda(0)C_\lambda(0). \tag{15.13}$$

Eqs. (15.10)-(15.11) lead to

$$A_\lambda(\phi)D_\lambda(\phi) = A_\lambda(0)D_\lambda(0). \tag{15.14}$$

Then,

$$\frac{C_\lambda(\phi)}{D_\lambda(\phi)} = \frac{C_\lambda(0)}{D_\lambda(0)}, \tag{15.15}$$

i.e.,

$$C_\lambda(\phi)D_\lambda(\phi) = C_\lambda(0)D_\lambda(0) \tag{15.16}$$

because $D_\lambda(\phi)$ and $D_\lambda(0)$ are ± 1, equal to their D_λ^{-1}. Then from Eq. (15.12) and Eq. (15.16), it follows that

$$A_\lambda(2\phi)B_\lambda(0)C_\lambda(0)D_\lambda(0) = -1 \tag{15.17}$$

which together with Eq. (15.9) gives

$$A_\lambda(2\phi) = A_\lambda(0) = \text{const}, \forall \phi. \tag{15.18}$$

But this cannot be valid if Eq. (15.8) is true. We must have

$$A_\lambda(\phi + \pi)B_\lambda(0)C_\lambda(\phi)D_\lambda(0) = 1. \tag{15.19}$$

Since from Eq. (15.10)

$$A_\lambda(\phi)B_\lambda(0)C_\lambda(\phi)D_\lambda(0) = -1, \tag{15.20}$$

then, dividing Eq. (15.19) by Eq. (15.20), we have

$$A_\lambda(\phi + \pi) = -A_\lambda(\phi) \tag{15.21}$$

which contradicts Eq. (15.18)!

In this way, we proved the impossibility of such a correlation in standard probability theory. This proof is also valid for three particles because the argument of B_λ is fixed as $\phi = 0$ in all our formulas.

15.2 HARDY'S EXAMPLE

Hardy gave another example [2,3] of a *quantum miracle* resulting from a correlation which is also totally impossible according to any classical reasoning. Let us discuss Hardy's example following Mermin[4].[1]

Two particles originate from some common source and fly in opposite directions to apparatuses on the left and right.

On the left apparatus it is possible to obtain the answer to one of the following two *yes-no* questions, A_l or B_l, and on the right apparatus, it is possible to obtain *yes-no* answers to two questions M_r or N_r.

Now, it is possible to prepare the two particles state $|\psi\rangle$, in such a way that the following results of the measurements on the right and left apparatuses are valid.

(a) If the questions are B_l and N_r, the answers are sometimes both *yes*.

(b) If the questions are either B_l and M_r or A_l and N_r, the answers are *never* both *yes*.

(c) If the questions are A_l and M_r, the answers are never both *no*.

The corresponding state vector which is able to satisfy the above requirements is a superposition of tensor products of one particle states $|A_l y\rangle, |A_l n\rangle, |M_r y\rangle, |M_r n\rangle$, where y, n mean *yes, no* answers: We write

$$|\psi\rangle = \alpha |A_l y\rangle \otimes |M_r n\rangle + \beta |A_l n\rangle \otimes |M_r y\rangle + \gamma |A_l y\rangle \otimes |M_r y\rangle. \tag{15.22}$$

Property (c) is fulfilled due to absence of *no, no* state $|A_l n\rangle \otimes |M_r n\rangle$ in superposition given by Eq. (15.23). In order to satisfy (a) and (b), it is necessary that $|\psi\rangle$ be orthogonal to $|B_l y\rangle \otimes |M_r y\rangle$ and to $|A_l y\rangle \otimes |N_r y\rangle$, i.e.

$$0 = \langle B_l y| \otimes \langle M_r y|\psi\rangle = \beta \langle B_l y|A_l n\rangle + \gamma \langle B_l y|A_l y\rangle \tag{15.23}$$

$$0 = \langle A_l y| \otimes \langle N_r y|\psi\rangle = \alpha \langle N_r y|M_r n\rangle + \gamma \langle N_r y|M_r y\rangle. \tag{15.24}$$

Now there is the important property (a), requiring that $|\psi\rangle$ is not orthogonal to $|B_l y\rangle \otimes |N_r y\rangle$. We have,

$$p = |\langle B_l y, N_r y|\psi\rangle|^2 \neq 0. \tag{15.25}$$

From Eqs. (15.23)-(15.24) it follows that

$$p = |\gamma^2| |\langle B_l y|A_l y\rangle|^2 |\langle N_r y|M_r y\rangle|^2. \tag{15.26}$$

Then, from normalization $|\alpha|^2 + |\beta|^2 + |\gamma|^2 = 1$ and Eqs. (15.23)-(15.24) we get

$$p = \frac{p_l(1-p_l)p_r(1-p_r)}{1-p_l p_r} \tag{15.27}$$

with

$$p_l = |\langle B_l y|A_l y\rangle|^2, \quad p_r = |\langle N_r y|M_r y\rangle|^2 \tag{15.28}$$

so that maximizing Eq. (15.27)) we get $p_l = p_r = (\sqrt{5}-1)/2$ and the maximal value of the *quantum miracle* is $p = 0.09017$. Now, let us show that what we predicted here is some *miracle* from the classical point of view of "existence" of properties of our particles.

Indeed, let us concentrate our attention on the properties (a), (b), (c), thinking that they *exist* as some *objective* properties and that there is *locality*. Then, the properties on the left are not changed if something different is measured on the right, and we come to a contradiction! Indeed,

(i) from (a) we look for the situation when both B_l and N_r give the answer *yes*.

Doing many experiments, we can concentrate on the particular runs for which (a) is fulfilled,

(ii) but *if* in these particular runs we ask other questions B_l and M_r, since on the left the answer is *yes*, on the right, due to property (b) the answer to M_r must be *no* .

(iii) If instead we ask for the values of A_l and N_r in the same run and the answer for N_r is *yes*, then the answer of A_r must be *no*.

(iv) If we ask for the values of A_l and M_r, from (ii) and (iii), it follows that the answer will be *no* both for A_l and to M_r questions. But this conclusion contradicts property (c)!

So our state vector really leads to something which is a *miracle* for a classically thinking man (woman). Nevertheless, this quantum result can be checked experimentally [4].

15.3 NEGATIVE EXPERIMENTS. ELITZUR-VAIDMAN PROPOSAL

When we discussed the quantum logic approach in Chapter 7, we said that *yes-no* values are given to quantum observables by a given observer. So, if there is, e.g., some *no* result, it means that a collapse of the wave function (or its preparation) took place.

Renninger [7] proposed a *negative* experiment; wave packet collapse occurs, but there is no physical interaction in the sense of energy transfer to some measuring apparatus. Imagine a source of electrons inside a spherical screen which has a hole, and such that the screen can oscillate if a particle arrives at it.

Now, let us investigate the situation when an electron leaves the source with some definite momentum (registered for example by checking some photon emitted by a source), but *no* oscillation of the screen due to the electron occurs, which means that it has gone through the hole. The wave function for the electron has then collapsed, so that it has a definite coordinate inside the hole, despite the lack of any energetic interaction between electron and the screen.

Another *quantum miracle* has been given by Elitzur and Vaidman[8] (henceforth cited as *EV*). Consider the interferometer of Fig. (15.1) with two arms.

A photon arrives at point 1, where there is a semi-transparent mirror. It can arrive at point 3 either by going through paths 123 or 143. If there is no possibility of checking which path the photon has followed, we have interference at point 3. Let this interference be such that detector D_2 near to point 3 will always give the answer *yes*, and detector D_1 will always answer *no*.

But, if somehow we *can* know what path is taken by the photon, the interference is broken and sometimes answer *yes* is possible for detector D_1. Now imagine that somebody has two kinds of bombs; the first kind *can* blow up if a photon arrives at its surface and the the second kind never explodes if photons arrive at its surface.

Can we guess by experimenting with photons, if a given bomb will explode, *without* exploding it? For classical reasoning, this is an impossible task, but let us see what quantum reasoning says.

Put the bomb at a point along the arm 143 .

If it is of the first kind, then taking into account that for a quantum particle the very *possibility* of explosion is important (because if the bomb *can* explode we can *know*

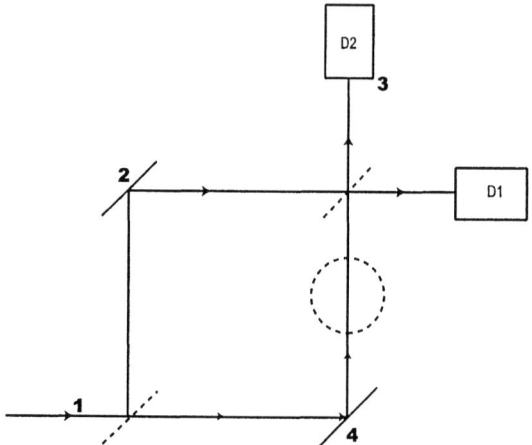

Figure 15.1. Mach-Zender Interferometer for the EV Experiment.

the path of the photon in the interferometer), the interference disappears. If the bomb is of the second kind, there is no interaction of the photon with it and we still have interference. Thus D_1 always answers *no*.

But there is an interesting possibility of a *negative* experiment, when the bomb in 143 can explode but it does not, because the photon chooses path 123! Then detector D_1 will give answer yes. The interference is destroyed and we will know that the weapon will explode if touched by a photon, without exploding it!

15.4 DELAYED CHOICE EXPERIMENTS

Von Weizsäcker, Bohr and Wheeler[9]2 proposed the so called *delayed choice experiment* in which the *past* seems to be defined by a future experiment. The idea is the following. We decide to activate a device that is able to detect a photon in some arm of an interferometer only at a time when the wave packet is already inside the interferometer. If there is no *influence on the past* due to wave packet collapse, which would be the case if the collapse is some physical process, then there could be situations in which we could see interference and at the same time fix which arm the photon has chosen. In order to be free from contradiction Bohr predicted that it does not matter "whether our plans for handling the instruments are fixed beforehand or whether we postpone the completion of our planning until a later moment when the particle is already on its way from one instrument to another" (see Schilpp[11]).

Wheeler proposed experiments carried out in which the choice is random either to measure or not in what arm the photon goes to (or in some experiments to *erase* information in what arm it went)[9]. Also, in quantum optics the so called *erasers* were proposed, in which Wheeler's delayed choice can be made.

These experiments confirm quantum theory according to which in *some sense* there is an influence of the future measurement on the past behavior of photons. Nevertheless, these experiments do not show influence of *future events* in Minkowski spacetime on the *past events* and do not mean nonlocality in time (contrary to the so-called time machines of general relativity).

This *past behavior of the photons* is really only some description but not a measurement of the past behavior in terms of *events* in the past. So causality is not disrupted.

Delayed choice experiments can lead to interesting problems in cosmology. Like Wheeler we might ask: "if we get information on the beginning of the universe (supposed to have happened at the Big Bang) through primordial radiation, which is a weaker radiation, the closer we get to the Beginning, what is the consequence of using quantum theory in this case?"

Really, coming over such a great distance, radiation is so weak, that quantum properties become important. But then, depending on the *choice* of the observer, what complementary observable he decides to measure, he could arrive at different conclusions about the *quantum beginning* of the universe, and in this way not only one, but many different beginnings can arise!

15.5 QUANTUM ERASERS

The most spectacular example of delayed choice is the *quantum eraser*, proposed by B. Englert, M. O. Scully and H. Walther[12,13]. Atoms are sent one by one through an apparatus consisting of two combined cavities—upper and lower. Then, as in the double slit experiment, they pass through a plate with two slits and arrive at the screen where spots are registered. Before arriving at the cavities, the atoms are excited by a laser beam. Figure 15.2 taken from[12] illustrates the experiment.

The upper and lower cavities are separated by two electrooptical shutters which can be on choice, either shut or opened. Between the shutters there is a thin-film semiconductor acting as a photodetector. The experiment starts with the cavities empty and the shutters closed. Atoms are excited by the laser light and then go to the cavities where they come to the ground state by spontaneous emission of a photon, which is left in one of the resonant cavities. If we could detect the photon, we could *know* which way the atom moved—through the upper or lower cavity! The wave function of the atom plus a photon can be written as

$$|\Psi\rangle = \frac{1}{\sqrt{2}}[\Psi_1(\vec{r})|1_1 0_2> + \Psi_2(\vec{r})|0_1 1_2>] \otimes |b>. \qquad (15.29)$$

Here $\vec{r}$ is the center of mass degree of freedom of the atom, $|b>$ denotes the final state of the atom, indices 1, 2 mean cavities 1 or 2, capital 1 and 0 with indices mean one photon in the corresponding cavity with no photon in the other. If nothing is done with shutters the probability density for atoms on the screen will be

$$P(\vec{R}) = \langle \Psi|\Psi\rangle = \frac{1}{2}[|\Psi_1|^2 + |\Psi_2|^2], \qquad (15.30)$$

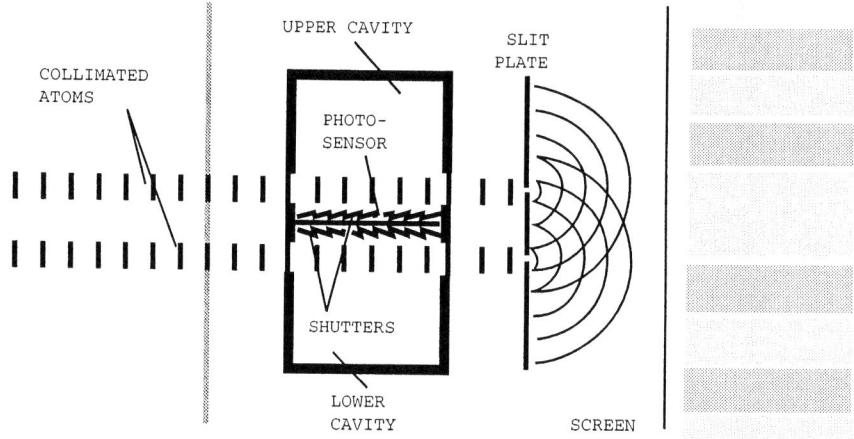

Figure 15.2. The Quantum Eraser. When an atom hits the screen, the shutters are opened. The spots on the screen are marked light or dark depending on if the sensor absorbs or not the photon in the cavity. Light spots produce an interference pattern and the dark generate a complementary pattern. (Adapted from a figure originally appearing in *Sci. American* **27**(6), 56–61 (1994) with permission from M. Goodman.)

since $< 1_1 0_2 | 0_1 1_2 > = 0$.

But now, include the detector in its ground state in our system. Then the wave function of the whole system will be

$$|\widehat{\Psi}\rangle = \frac{1}{\sqrt{2}}[\Psi_1(\vec{r})|1_1 0_2 > + \Psi_2(\vec{r})|0_1 1_2 >]|b> \otimes |d>. \qquad (15.31)$$

Absorbing a photon, the detector would be found in an excited state "e". Changing basis:

$$\Psi_\pm = \frac{1}{\sqrt{2}}[\Psi_1(\vec{r}) \pm \Psi_2(\vec{r})],$$

$$|\pm> = \frac{1}{\sqrt{2}}[|1_1 0_2> \pm |0_1 1_2>], \qquad (15.32)$$

we can write

$$|\widehat{\Psi}\rangle = \frac{1}{\sqrt{2}}[\Psi_+(\vec{r})|+> + \Psi_-(\vec{r})|->]|b> \otimes |d>. \qquad (15.33)$$

Now we come to the crucial point. Imagine that our atom, emitting the photon, *passed* through the double slit plate and is far from our cavities. Then open the shutters. Before opening the shutters we could get "which way" information, so no interference on the screen could be seen. But by opening them we can *erase* this information, delaying as long as we want our choice to *erase* it or not! If the detector registers the photon we cannot say in which cavity the photon was, together with the atom emitting it. So interference must reappear! Suppose that our detector can be excited only by a photon in the symmetric state, going from state $|d>$ to $|e>$. So the detector's action produces the state

$$|\Phi\rangle = \frac{1}{\sqrt{2}}[\Psi_+(\vec{r})|+>\otimes|e> + \Psi_-(\vec{r})|->\otimes|d>]\otimes|b>. \quad (15.34)$$

Again on the screen we have the probability density

$$P(\vec{R}) = \frac{1}{2}(|\Psi_1(\vec{R})|^2 + |\Psi_2(\vec{R})|^2), \quad (15.35)$$

where $\vec{R}$ is the coordinate on the screen. But if we ask about the probability density of finding both the excited detector and the atom on the screen, we obtain

$$P_e(\vec{R}) = |\Psi_+(\vec{R})|^2 = \frac{1}{2}[|\Psi_1(\vec{R})|^2 + |\Psi_1(\vec{R})|^2 + Re(\Psi_1^*(\vec{R})\Psi_2(\vec{R}))]. \quad (15.36)$$

So, we shall see interference for those atoms which correspond to photons that excited the detector. Now, the probability density for finding the detector deexcited and the atom on the screen (with $\vec{R}$ being its position) is

$$P_d(\vec{R}) = |\Psi_-(\vec{R})|^2 = \frac{1}{2}[|\Psi_1(\vec{R})|^2 + |\Psi_2(\vec{R})|^2 - Re(\Psi_1^*(\vec{R})\Psi_2(\vec{R}))]. \quad (15.37)$$

So if the *eraser photon signal* is disregarded, we do not see interference. But if we somehow *label* our atoms according to the detector being excited or not (*red* and *green* in the example of Englert et al.), we can see interference patterns for the labeled atoms. So here there is nonlocal correlation between the interference pattern on the screen and the far distant from it excitations of the detector! In their paper Englert et al. go even farther. They say: "As the photon remains in one of the cavities, the atom reaches the screen where it leaves a spot. Once that happened we open the shutters simultaneously, turning the two separate cavities into a single larger one." Thus, *after* the atom was checked on the screen, the photosensor *detected* the photon, in spite of the fact that it is usually thought that *because* the photon is detected we have interference ... Here we come to the very controversial problem concerning the influence of the future on the past. It was the *future* choice to erase information that *influenced* the atoms to give an interference pattern in the past!

This possibility has been criticized by V. Mohrhoff[14]. The author agrees that the interference pattern will appear if the delayed choice is made *before* atoms come on the screen. This choice can be made when atoms are far from the cavities but still are not registered as spots on the screen. Nevertheless, he claims that we cannot see the interference *after* atoms are registered, saying that the detection of photons will change only the states of the photons themselves, without *labeling* the atoms. However, we can comment on this situation by comparing it with the typical *EPR* situation, discussed by us previously. If we agree that in some reference frame the choice is made *before* registration, there exists, for spacelike intervals, another frame in which it is made *after* it. So, if labeling is possible for one time ordering, it must be possible for the other ordering, as time *labeling* is not relativistically invariant for space like events. Surely we cannot get information from the future due to the *erasure effect*. We must know both groups of events—spots of atoms on the screen and excitations of the detector—in order to *select* the interference pattern. Without knowledge of what is checked in the future, it is impossible to say anything about that future. As Englert et al. say, "a black-and-white photograph of the screen would not show the interference pattern." Differently from the conjectured *time machines* of general relativity, where some events (like killing one's grandfather) are prohibited by influence from the future, here no events are prohibited. But the statistics of some events is conditioned by future, as if the future puts its *shadow* on them ... It would be very interesting to see the experimental results for different time ordered delayed choice and registration of atoms on the screen in the situation discussed here!

15.6 QUANTUM CRYPTOGRAPHY

The fact that a measurement changes the wave function can be useful in cryptography,[3] as discussed by Deutsch[16], Bennett et al[17] and Ekert et al[18]. *Cryptography* concerns the conservation of *secrets* in communications between persons and it is important in politics and economy. Here we discuss, following Ekert et al[18], how to produce a cryptogram by taking into account the breakdown of Bell's inequalities and the nonlocality presented in *EPR* correlated pairs. After that we discuss that *unfortunately*, quantum cryptography does not permit secure bit commitment.

Imagine two people, Alice and Bob, who want to send each other secret communications. Imagine also, that a third person, Eve, tries to find their secrets out. In classical physics Eve can always (without visual *disturbance*) intervene in the classical *channel* of communication (for example, by receiving electromagnetic waves) so that Alice and Bob will not know that she knows the secret key used in the message.

It is not so in quantum physics. If Alice and Bob use a *quantum channel*, which is a source emitting pairs of spin 1/2 particles in a singlet state moving along the z-axis in the direction of the users, they measure the spin projections $\vec{a}_i, \vec{b}_j$ ($i, j = 1, 2, 3$) respectively lying in the xy plane, characterized by the azimuthal angles: $\phi_1^A = 0$, $\phi_2^A = \pi/4$, $\phi_3^A = \pi/2$ and $\phi_1^B = \pi/4$, $\phi_2^B = \pi/2$, $\phi_3^B = 3\pi/4$.

Alice and Bob choose the orientation of the analyzers that measure the spin projection randomly and independently for each pair. Each measurement gives result ± 1.

The correlation coefficient of the measurements is given by

$$E(\vec{a}_i, \vec{b}_j) = P_{++}(\vec{a}_i, \vec{b}_j) + P_{--}(\vec{a}_i, \vec{b}_j) - P_{+-}(\vec{a}_i, \vec{b}_j) - P_{-+}(\vec{a}_i, \vec{b}_j) \quad (15.38)$$

where $P_{\pm\pm}(\vec{a}_i, \vec{b}_j)$ are the probabilities of results ± 1 along $\vec{a}_i$ and ± 1 along $\vec{b}_j$. We already know (see Chapter 4), that

$$E(\vec{a}_i, \vec{b}_j) = -\vec{a}_i \cdot \vec{b}_j. \tag{15.39}$$

So, for the same orientation $\vec{a}_2, \vec{b}_1$, and $\vec{a}_3, \vec{b}_2$, we have total anticorrelation of the results obtained by Alice and Bob.

$$E(\vec{a}_2, \vec{b}_1) = E(\vec{a}_3, \vec{b}_2) = -1. \tag{15.40}$$

Now, following Bell, Clauser, Horne, Shimony and Holt, define

$$S = E(\vec{a}_i, \vec{b}_1) - E(\vec{a}_1, \vec{b}_3) + E(\vec{a}_3, \vec{b}_1) + E(\vec{a}_2, \vec{b}_3) \tag{15.41}$$

which according to quantum theory has the value

$$S = -2\sqrt{2}. \tag{15.42}$$

After transmission of the pairs of particles, Alice and Bob announce publicly the orientations of the analyzers chosen and divide the results of measurements on two groups: a) when they used a different orientation for the analyzers; b) when the orientations are the same.

In public communication through a classical channel (for example by radio) they first announce the results obtained, but only for group a), and then calculate S in Eq. (15.41). If Eve intervened in the *quantum channel*, her measurement certainly would spoil the entangled state vector of the two particles and as a consequence Bell's estimate would be valid, and

$$-\sqrt{2} \leq S \leq \sqrt{2}, \tag{15.43}$$

contrary to Eq. (15.42).

If Alice and Bob calculate S and find the result given by Eq. (15.42) they are sure that nobody (Eve) intervened in their communication. But then, they can use the results in the second (secret) group of anticorrelated measurements, made by them randomly, as a secret string of bits encoding the secret key, which is known to them and to nobody else.

15.6.1 QUANTUM BIT COMMITMENT

The above ideas which are technically known as *quantum key distribution (QKD)* lead for some time to the hope that quantum theory opens the door for new, practical, and safe communication systems. However, to know if someone read or not the message that Alice sent to Bob, is not the unique proposal of cryptography. More important is the problem called *bit commitment*. This consists in the following: Suppose Alice put into a safe a bit b that she wrote on a piece of paper and sends it, without the key, to Bob. The objective of Alice and Bob's affair and in bit commitment in general, is that Alice cannot change her mind about the value of bit b and Bob, on the other hand, cannot determine bit b, until a late time when Alice decides to unveil b by giving the key to Bob. In 92, it has been proposed by Bennett and Brassard[19] to use quantum theory to produce an absolutely secure *quantum bit commitment (QBC)* protocol. In that paper the authors noted at the end that their protocol was in fact insecure, since (see below) in their scheme it was always possible for Alice to cheat by using *EPR* pairs (e.g., of photons) when sending the commitment to Bob. In 95, Brassard et al[20] proposed a new protocol for *QBC* which they thought to be unconditionally secure and this has been accepted for some time[21]. However, as recently proved by Lo and Chau[22] and also by Mayers [23], an unconditionally safe *QBC* protocol is impossible. In brief, the same *EPR* phenomenon that permits secure transmission of information in the case of the *QKD* as described above, can be used by Alice in *QBC* to cheat Bob when she wants. In the following, we present a proof of the above statement.

To start, let us put the problem as originally imagined by Bennet and Brassard[19]. The first step in the protocol is known as *commitment phase*. It consists of:

(a) A bit b with values 0 *or* 1 that Alice wants to commit to Bob is encoded in a quantum system, let us say, in a sequence of s (s is called the *security parameter*) polarized photons that are sent to Bob.

The sequence of the s photons is prepared by Alice randomly on one of two different polarization basis, i.e., *each one of them* is (randomly) chosen as,

(i) being horizontally or vertically polarized (rectilinear basis) if $b = 0$.

(ii) being 45° or 135° polarized (diagonal basis) if $b = 1$.

Of course, in both case Alice takes *note* of the sequence, and keeps it *secret*.

(b) Bob receives the s photons and randomly measures the polarization of each one of them in the rectilinear or diagonal basis, and, of course, takes note of the sequence.

It is clear that, on the average, Bob detects only $s/2$ photons in the original basis chosen by Alice. Anyway, in an ideal case[4], his ignorance with respect to the value 0 or 1 chosen by Alice is total since a simple calculation (c.r., see Eq. (2.16)) shows that the density matrices corresponding to these two cases are equal and proportional to the unity operator.

The second step in the protocol is called the *opening phase,* when Alice opens her commitment declaring to Bob the value of b that she has chosen. Of course, at that time she also gives Bob the notes she took of the polarization of the sequence of the s photons.

Since as said above, Bob on the average detected $s/2$ photons in the original polarization basis, he can verify for these cases, if Alice's notes match his own notes.

Observe, that it is almost impossible for Alice to cheat under the conditions of the above experiment, since the maximum that she could do, e.g., in the case she has chosen $b = 0$ in the commitment phase and wants to declare $b = 1$ in the opening phase, is try to guess the polarizations along the diagonal basis detected by Bob. It is obvious that the probability for cheating with success is $(1/2)^{\frac{s}{2}}$, being very small with a large s.

But now, imagine that Alice sends Bob the sequence of s photons such that each one of them is a member of an *EPR* pair. In this case she can cheat with success if she wants. Indeed, all that she needs to do is to skip her measurements during the commitment phase and to decide on the value of b only at the opening phase. For $b = 0$, she measures the polarizations along the rectilinear basis and for $b = 1$, she measures along the diagonal basis. Due to the *EPR* correlation, for each one of Alice's measurements of the polarization of her sequence of the s stored photons, she will obtain a sequence of polarizations that is always orthogonal to the sequence of polarizations obtained by Bob. Then Alice always can announce the convenient sequence to fulfill her intention without Bob's having a single way to detect that his is being cheated on.

Is there any possibility of Bob's preventing this kind of cheating? The answer is *no*.

Recall that in all QBC protocols, it is necessary that Alice sends Bob two kinds of information, one in the commitment phase (the sequence of s photons) and one in the opening phase when she declares the value of b and her notes about the polarizations of each one of the s photons. This last step, can be done by a classical communication channel, but from the theoretical point of view it also can be done using a quantum channel. We have then, the following general procedure based on *five steps* for a two party QBC [22,23], which is to be executed in a system modeled by state vectors in a Hilbert space

$$\mathcal{H} = \mathcal{H}_A \otimes \mathcal{H}_B, \tag{15.44}$$

where $\mathcal{H}_A$ and $\mathcal{H}_B$ are respectively the Hilbert spaces which correspond to the effective systems employed by Alice and Bob when executing the protocol. In what follows, $\mathcal{H}_A$ and $\mathcal{H}_B$ are supposed to be two dimensional quantum registers.

First step. We analyze the ideal case first. In the commitment phase Alice chooses the value of b. She prepares for, depending on her choice $b = 0$ or $b = 1$, the following quantum states (which are supposed to be known both by Alice and Bob),

$$|0\rangle = \sum_k \lambda_k |e_k\rangle_A \otimes |\phi_k\rangle_B,$$

$$|1\rangle = \sum_k \lambda_k |e'_k\rangle_A \otimes |\phi_k\rangle_B. \tag{15.45}$$

where $\{|e_k\rangle_A\}, \{|e'_k\rangle_A\}$ are two orthonormal bases for $\mathcal{H}_A$ and $\{|\phi_k\rangle_B\}$ is an orthonormal basis for $\mathcal{H}_B$ and the λ_k are the eigenvalues of each one of the following two of the identical reduced density matrices (see section 2.2),

$$\rho_B^0 = Tr^{(A)}|0\rangle\langle 0|, \quad \rho_B^1 = Tr^{(A)}|1\rangle\langle 1|$$
$$\rho_B^0 = \rho_B^1 \tag{15.46}$$

which describes the state that Bob will receive after Alice's measurement.

Alice's choice of $|0\rangle$ and $|1\rangle$ then assures that Bob will not know her choice. As we shall see, this will permit Alice to cheat, always.

Second step. Alice, if honest, will make a measurement on her quantum register deciding on the value of b, which implies that she will produce a collapsed quantum state that is either described by ρ_B^0 or ρ_B^1, depending on her choice.

Third step. Alice sends information for Bob's register as evidence for his commitment.

Fourth step. At the opening phase, Alice declares her commitment, i.e., the value of b. If $b = 0$ (respectively, $b = 1$), she will inform that in the second step she made her measurement using the basis $\{|e_k\rangle_A\}$ (respectively, $\{|e'_k\rangle_A\}$).

Fifth step. Bob made measurements on his register to verify if Alice has indeed committed to a genuine bit.

This is done, of course, by comparing (correlating) Alice's information obtained in the fourth step with his own results. If the expected correlations show up, Bob will accept that Alice made an honest protocol, and if not he can be sure that Alice was cheating.

Well, the fact is that Alice can *always cheat*. This happens because it is *possible* that,

(i) Alice, not being honest, will postpone her measurement, supposed to be done in the second step of the commitment phase, until the beginning of the opening phase[5].

(ii) Then, if she chose in the beginning $b = 0$ and changed her mind and wants to declare that she chose $b = 1$, during the opening phase she simply rotates the prepared quantum state $|0\rangle$ into $|1\rangle$ in $\mathcal{H} = \mathcal{H}_A \otimes \mathcal{H}_B$ with physical operations carried out *only* in her quantum register, i.e., without Bob's help. Mathematically, this is realized with a unitary transformation

$$\mathbf{U} = \mathbf{U}_A \otimes \mathbf{1} \in \mathcal{H} \otimes \mathcal{H}^\dagger, \qquad (15.47)$$

where

$$\mathcal{H}_A \otimes \mathcal{H}_A^\dagger \ni \mathbf{U}_A : \{|e_k\rangle_A\} \mapsto \{|e'_k\rangle_A\}). \qquad (15.48)$$

It is the possibility that Alice has of *skipping* the second step in the protocol, that invalidates Brassard et al[20] argument.

It remains to be proven, that even for the non ideal case, where the reduced density matrices ρ_B^0 and ρ_B^1 corresponding to Alice's two possible choices are *different* but differ in a small amount (called closeness), that cheating is possible[23]. To define the closeness between two reduced density matrices, we first introduce the concept *purifications* of a density matrix. For the density matrices ρ_B^0 and ρ_B^1, this means that there exist pure states $|\Phi_0\rangle, |\Phi_1\rangle \in \mathcal{H} = \mathcal{H}_A \otimes \mathcal{H}_B$, such that

$$\rho_B^0(|\Phi_0\rangle) = Tr^{(A)}|\Phi_0\rangle\langle\Phi_0|, \quad \rho_B^1(|\Phi_1\rangle) = Tr^{(A)}|\Phi_1\rangle\langle\Phi_1|$$
$$\rho_B^0(|\Phi_0\rangle) \neq \rho_B^1(|\Phi_1\rangle). \qquad (15.49)$$

Second, we introduce the concept of *fidelity* of two reduced density matrices obtained through two purifications, $|\Phi_0\rangle$ and $|\Phi_1\rangle$ by the function

$$F[\rho_B^0(|\Phi_0\rangle), \rho_B^1(|\Phi_1\rangle)] = \begin{cases} 1 & \text{if } \rho_B^0(|\Phi_0\rangle) = \rho_B^1(|\Phi_1\rangle) \\ \max|\langle\Phi_0|\Phi_1\rangle| & \text{if } \rho_B^0(|\Phi_0\rangle) \neq \rho_B^1(|\Phi_1\rangle) \end{cases} \qquad (15.50)$$

where the maximization is over all possible purifications.

We observe that it is obvious that $\rho_B^0(|\Phi_0\rangle)$ cannot differ significantly from $\rho_B^1(|\Phi_1\rangle)$, since in this case the protocol is not safe because now it is Bob that can cheat. We then suppose that the fidelity differs only by a small amount from 1. To see how Alice can cheat also in this case, it is enough to observe, e.g., that if $|\Phi_1\rangle = |1\rangle$ there always exists a $|\Phi_0\rangle$ such that

$$F[\rho_B^0(|\Phi_0\rangle), \rho_B^1(|1\rangle)] = |\langle\Phi_0|1\rangle| = 1 - \delta, \quad \delta > 0. \qquad (15.51)$$

If Alice, having chosen in the commitment phase $b = 0$ and thus prepared the state $|0\rangle$, skips the second step of the protocol and decides at the opening phase to change her mind and declare that $b = 1$, all she needs to do is to rotate the state in her register from $|0\rangle$ to the state $|\Phi_0\rangle$ such that Eq. (15.51) holds. As in the ideal case, this is always possible without Bob's knowing what happened, since it involves only a unitary transformation like in Eq. (15.47). This completes the proof that safe *QBC* is impossible.

15.7 TELEPORTATION

EPR correlations can be used for realization of the interesting possibility of *teleporting* the unknown quantum state of some particle at a given place to another particle in some other place. The idea of an experiment to verify this extraordinary prediction of quantum theory was done in 1993 by Bennett et al[24], and the experiment itself was recently performed by a group of Innsbruck University headed by Zeilinger[25] and at the same time by a group of Rome University headed by De Martini[26]. Interesting comments on these experiments have been made by Collins[27].

These experiments, together with Aspect experiments[28,29] and the recent CERN experiment done by Tittel[30] showing nonlocal correlations for distances of the order of 10 km can be considered as one more confirmation of the basic principles of quantum physics.

The name *teleportation* first appeared in science fiction as the name for the possibility to convert a man into a pulse of electromagnetic radiation, send him to some distant planet in another galaxy, where highly civilized creatures can reconstruct him using atoms from their planet, following instructions coded in the pulse. If they follow these instructions carefully, they will reconstruct the man with his brain and memory stored in it, so that the whole affair will seem like a trip of a man to that planet. This *teleportation* is different from *cloning* of the same man, because the man has disappeared from Earth. In the case of *cloning* a *copy* of the man appears on the other planet while the original man still exists on Earth. As we shall see, *cloning* is *impossible* in quantum physics while *teleportation* is *possible*.

Suppose, there are two observers, Alice and Bob, located in two different places in space. Now, suppose that Alice has a given quantum system such as a photon or spin 1/2 particle, prepared in a state unknown to her, and that she wishes to communicate sufficient *information* to Bob about the quantum system for him to make an exact copy of it. Surely, if Alice knew the state vector of her particle, she could inform Bob about that state using a classical channel of communication, e.g., using radio waves. But, by hypothesis she does not know the state vector of her quantum system, and if she measures some observable of the system, the original state will be destroyed due to wave packet collapse, and as a result, the system will change (in general) its state vector. Only if somebody told Alice beforehand that the state vector of the quantum system had been prepared as an eigenvector of some observable is that an eventual measurement made by Alice of the same observable would permit the exact knowledge of the state vector. In what follows, we do not consider this case.

Surely, she could send the particle itself to Bob, trying to do it without changing the unknown state. But the particle could be unstable and then, it could be impossible to send it to Bob because of its strong interaction with the environment, or Alice simply would not know Bob's location! Nevertheless quantum theory and its nonlocality give an interesting possibility noticed in [24] of using (like in quantum cryptography) two channels, one classical and one quantum (of *EPR* type) for teleportation. The idea is the following. Alice makes her particle interact unitarily with another quantum system, the so called *ancilla*, initially in a known state, constituting an *EPR* pair. One of the particles from the *EPR* pair goes to Bob, the other to Alice. Then, she makes some measurement on the two particles system composed of her particle in the unknown state and her *EPR* particle. She communicates the result of her measurement to Bob by using any classical channel and the surprise is that this is enough for Bob to do some operation after which it is warranted that his particle acquires a quantum state vector that is the same state as the one possessed by Alice's original particle. The example of [24] corresponds to *spin-exchange measurement* in which the state is totally characterized by the spin or polarization variable so that what is *teleported* is polarization. The mathematical description of teleportation in this case is as follows. Suppose we have two spin 1/2 particles called 2 and 3 prepared in an *EPR* singlet state, which using a slight modified notation in relation to Eq. (2.14) and Eq. (7.29) is written as

$$|\Psi_{23}^{(-)}\rangle = \sqrt{\frac{1}{2}}(|z,\uparrow\rangle_2 \otimes |z,\downarrow\rangle_3 - |z,\downarrow\rangle_2 \otimes |z,\uparrow\rangle_3) \in \mathcal{H}_2 \otimes \mathcal{H}_3. \quad (15.52)$$

Here, the *subscripts* 2 and 3 label the particles of the *EPR* pair. Alice also has particle 1, which can be of a kind different from 2 and 3 and is in a state unknown to her. The three particle system is initially in the state, described by $|\Psi_{123}\rangle = |\Phi_1\rangle \otimes |\Psi_{23}^{(-)}\rangle \in \mathcal{H}_1 \otimes \mathcal{H}_2 \otimes \mathcal{H}_3$.

The crucial point is when Alice measures the two particle joint system, consisting of particles 1 and 2!

The results of her measurements can be described by four functions of the so called *Bell's operator basis* (note that particles are not identical)[31] of $\mathcal{H}_1 \otimes \mathcal{H}_2$,

$$|\Psi_{12}^{(\pm)}\rangle = \sqrt{\frac{1}{2}}(|z,\uparrow\rangle_1 \otimes |z,\downarrow\rangle_2 \pm |z,\downarrow\rangle_1 \otimes |z,\uparrow\rangle_2),$$

$$|\Phi_{12}^{(\pm)}\rangle = \sqrt{\frac{1}{2}}(|z,\uparrow\rangle_1 \otimes |z,\uparrow\rangle_2 \pm |z,\downarrow\rangle_1 \otimes |z,\downarrow\rangle_2). \quad (15.53)$$

Write the unknown state of particle 1 as

$$|\phi_1\rangle = a|z,\uparrow\rangle_1 + b|z,\downarrow\rangle_1, \quad (15.54)$$

where $|a|^2 + |b|^2 = 1$. Then,

$$|\Psi_{123}\rangle = \frac{1}{2}[|\Psi_{12}^{(-)}\rangle \otimes (-a|z,\uparrow\rangle_3 - b|z,\downarrow\rangle_3) + |\Psi_{12}^{(+)}\rangle \otimes (-a|z,\uparrow\rangle_3 + b|z\downarrow\rangle_3)$$

$$+ |\Phi_{12}^{(-)}\rangle \otimes (a|z,\uparrow\rangle_3 + b|z,\downarrow\rangle_3) + |\Phi_{12}^{(+)}\rangle \otimes (a|z,\uparrow\rangle_3 - b|z,\downarrow\rangle_3)]. \quad (15.55)$$

It follows that Alice can obtain one of four different results for her measurement of the two particle system, all with the same probability 1/4. But, to a given result obtained by Alice, there corresponds some definite state for Bob's particle, so that after the wave packet collapse, which is a nonlocal phenomenon, his state becomes one of the four possibilities

$$|\phi_3\rangle \equiv \begin{pmatrix} a \\ b \end{pmatrix}, \quad \begin{pmatrix} -1 & 0 \\ 0 & 1 \end{pmatrix} |\phi_3\rangle, \quad \begin{pmatrix} 0 & 1 \\ 1 & 0 \end{pmatrix} |\phi_3\rangle \text{ and } \begin{pmatrix} 0 & -1 \\ 1 & 0 \end{pmatrix} |\phi_3\rangle. \quad (15.56)$$

Each of these states is definitively related to the original state of Alice's particle 1. In the case of the first outcome (apart from an irrelevant phase factor), if Bob got information from Alice about her result, he would know that he has a replica of the spin state of Alice's particle 1. In the other cases Bob must apply one of the unitary operators in Eq. (15.56), corresponding respectively to 180° rotation around the z, x and y axes in order to obtain a perfect replica of the spin state of Alice's particle 1. In any case, once Bob has gotten information from Alice, he immediately knows what to do in order to get the replica. For the photon's polarization, these unitary operations can be physically carried out by using combinations of half-wave plates.

So, as a result of the above experiment in which Alice performs a measurement on a two particle sub-system of the three particle system originally in the quantum state $|\Psi_{123}\rangle$ the original state of 1 is *destroyed* (that is why this process is not *cloning*) but the spin state of particle 1 appears instantaneously at the place where Bob is, this being the reason why the process is appropriately called *teleportation*.

Alice's measurement produces two bits of classical information, because from four equally probable possibilities one outcome is obtained and this information is transmitted to Bob. In order to eliminate misunderstandings, it is important, to never forget that *without* Alice's classical communication with Bob, it is impossible for him to reproduce the spin state of Alice's particle 1. Indeed, if this could eventually be possible, it would mean transmission of information with *superluminal velocity*. But, as we already know, this cannot be achieved in *EPR* experiments.

The simple scheme of the experiment involving three particles can be generalized for four particles when Alice's particle 1 in itself is a member of an *EPR* singlet with some particle denoted 0. After teleportation, particles 0 and 3 will be in the singlet state while initially, before Alice's measurement, they belonged to separate *EPR* pairs. Instead of the singlet state of the *EPR* pair we may also take some general entangled state on the basis of N orthogonal functions where Alice measures the appropriate observable for the two-particle system, projecting the state onto some[6]

$$|\Psi_{nm}\rangle = \sum_j e^{2\pi i j n / N} |z, j\rangle_0 \otimes |z, (j+m) \mod N\rangle_1 / \sqrt{N}. \quad (15.57)$$

If Bob gets from Alice the result nm, he performs the unitary operation

$$\mathbf{U}_{nm} = \sum_k e^{2\pi i k n / N} |k\rangle\langle (k+m) \mod N|, \quad (15.58)$$

on his previously entangled particle 3 and gets the original spin state of Alice's particle 1. In the Innsbruck experiment[30], pairs of photons were produced in an entangled state using the process of parametric down-conversion (see fig. 15.7). A pulse of ultraviolet radiation passing through a nonlinear crystal creates the *ancillary* pair of photons 2 and 3. After retroflection during its second passage through the crystal the ultraviolet pulse creates another pair of photons, one of which is to be prepared in the initial state of photon 1 to be teleported, the other one serving as a trigger, indicating that a photon to be teleported is under way.

Alice then looks for coincidences after the beam splitter BS, in which initial photon 1 and one of the *ancillas* form a two-particle quantum state.

Bob, after receiving the classical information that Alice obtained a coincidence count in detectors f_1 and f_2 identifying the singlet 12 Bell state, knows that his photon 3 is in the initial state of photon 1. He can check this fact by doing a polarization analysis with the polarizing beam splitter PBS and the detectors d_1 and d_2. Detector p informs that photon 1 is under way. First, teleportation was demonstrated for states from the complete basis of 45° and -45° linear polarizations. Second, it was shown to occur for circularly polarized photons, the states of which are superpositions of the previous ones. Also, polarizations 0° and 90° were considered.

For example, if photon 1 was polarized at 45° and a f_1 f_2 coincidence takes place (meaning that the system composed of photons 1, 2 are in a singlet state), then photon 3 should be polarized at 45°. This can be easily checked by passing it through a polarizing beam splitter to select needed polarizations. Teleportation means that only the detector d_2 at 45° output of the beamsplitter will click. So, a three–fold coincidence d_2 f_1 f_2 together with absence of a three-fold coincidence d_1 f_1 f_2 (−45° analysis) is a *proof* of teleportation of the state of photon 1 to photon 3. The experiment proved the reality of teleportation.

Nevertheless, we must agree with Collins[27] that "the Innsbruck Alice did not perform a complete Bell measurement. Instead, she distinguished only one of the Bell states from the other three. Thus teleportation could be achieved at best only in 25% of the time—on those occasions when Alice happened to detect." However visibilities of 75% were achieved for teleportation, while for a classical system, visibilities of only up to 50% could be achieved.

The group in Rome used another idea for teleportation. Instead of taking the system of four photons as the Innsbruck team did, it used a two-photon system forming an *EPR* pair, but each photon is characterized by two different degrees of freedom, its *polarization* and its *path*. Alice has four detectors to fix one of the four Bell states of these two degrees of freedom, and then communicates the result of her measurement to Bob. What is *teleported* to Bob is the *polarization state* which can be recovered by Bob (after receiving information through a classical channel from Alice), once he uses one of a set of four unitary transformations similar to the ones above. The group teleported linear and elliptically polarized states. The results *confirmed* the theory with visibilities of more than 80%. In other tests, a difference of more than 8 standard deviations in relation to the predictions of classical theory was achieved.

From the point of view of basic principles, quantum teleportation is interesting because it gives an example of a totally new possibility of *objectively existing logical implication*: the quantum state of a system in a given place is teleported to another particle in a distant place not due to some physical process like, e.g., the propagation

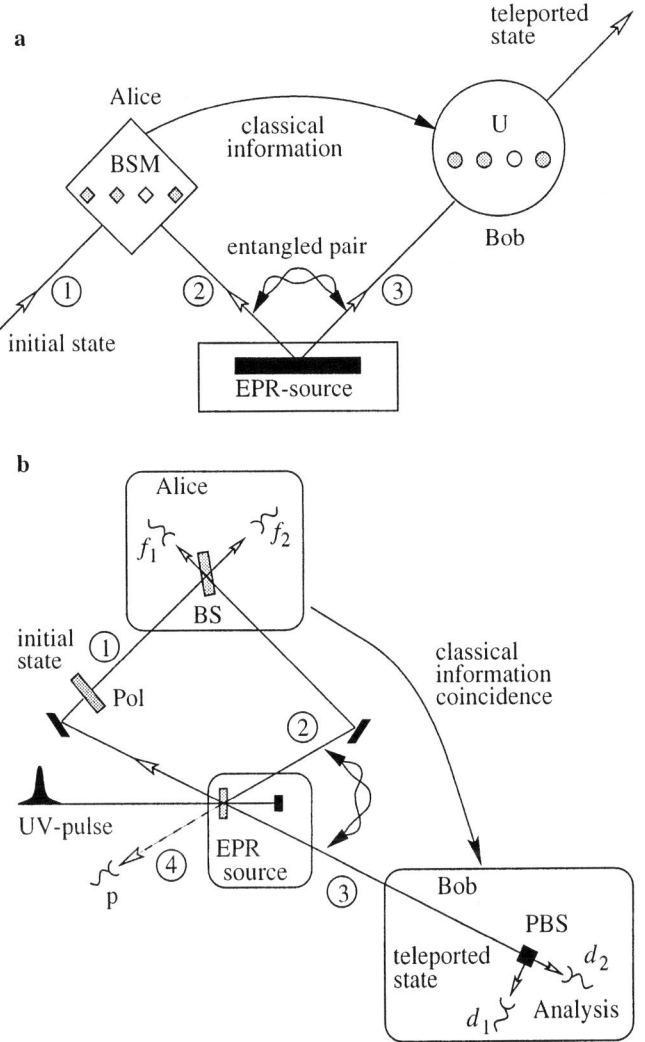

Figure 15.3. Scheme for the quantum teleporting experiment used by the Innsbruck group. (Reprinted with permission from *Nature*[25], Copyright (1997) Macmillan Magazines Limited)

of the particle or radiation from one place to another, but due to wave packet collapse. Alice's measurement *implies* that Bob's particle *acquires* a well defined quantum state.

The Innsbruck and Rome experiments *convincingly* illustrate some of the basic principles of quantum physics which are:

(i) the *projection postulate* which implies in the possibility for Alice to prepare nonlocally the teleported state which Bob receives,

(ii) the *postulate that the state vector really describes a single system* and *not* only an ensemble of identical systems. (Indeed, the experiment shows that teleportation can be made for individual photons).

We observe that the phenomenon of teleportation can be eventually useful in building quantum computers. Indeed, it could be the basis for projecting a *quantum memory* such that information from some domain (and encoded in some quantum state vector) be teleported through photons to another domain and there stored in the same quantum state.

Also, teleportation is useful for *EPR* swap-transferring the entanglement to Bob's particle, when his particle becomes entangled with the teleportees partner, "even though the two particles have neither interacted nor shared a common past. A test of Bell's theorem using entanglement swapping could test nonlocality with a pair of particles that have neither interacted directly." [27]

In resume, what can we say about teleportation? An appropriate answer has already been given by Aspect which according to Collins[27] said: "It would certainly help us to further understand nonlocality."

To end this Chapter, we must say that we regret leaving other interesting experiments out of our discussions, made or proposed in quantum optics and that concern nonlocality and measurements of momentum, energy and travel time for photons in barriers and wave guides (the so-called dwell time for tunneling effect), because different from the simple spin measurements done in Aspect's experiments, in these cases there are still controversial questions concerning: (i) the very meaning of the "uncertainty relations" $\Delta E \Delta t \geq \hbar$ which are different, as is well known, from the standard Heisenberg uncertainty relations; (ii) the meaning of what constitutes a signal, etc. A comprehensive discussion of these issues will require a book almost of the size of the present one, and is in preparation by one of the authors (W. A. R.).

Notes

1. Experiments testing Hardy's conjecture have been done with positive results[5,6]

2. A discussion of the delayed choise experiment and many references can be found in [10].

3. Quantum cryptography was started by Wiesner in a paper written in 1970 but which remained unpublished until 1983[15].

4. See below for the non ideal case where the *fidelity* of the density matrices differs by a small amount.

5. This will necessitate the use of a quantum computer, not yet available, but in principle possible of being built some day[22].

6. The indices 0 and 1 in the second member of Eq. (15.57) indicates that $|\Psi_{nm}\rangle \in \mathcal{H}_0 \otimes \mathcal{H}_1$.

References

[1] D. M. Greenberger, M. A. Horne and A. Zeilinger, in: *Bell's Theorem, Quantum Theory, and Conceptions of the Universe*, edited by M. Kafatos (Kluwer Academic Publishers, Dordrecht, 1989), pp.73–76.

[2] L. Hardy, Quantum mechanics, local realistic theories and Lorentz invariant realistic theories, *Phys. Rev. Lett.* **68**(20), 2981–2984 (1992).

[3] L. Hardy, Nonlocality for 2 particles without inequalities for almost all entangled states, *Phys. Rev. Lett.* **71**(11), 1665–1668 (1993).

[4] D. Mermin, What's wrong with this punctuation, *Phys. Today* **47**(11), 119–119 (1994).

[5] J. R. Torgerson, D. Branning, C. H. Mouken, L. Mandel, Experimental demonstration of the violation of local realism without Bell inequalities, *Phys. Lett. A* **204**(5-6), 323–328 (1995).

[6] G. DiGiuseppe, F. De Martini, D. Boschi, Experimental test of the violation of local realism in quantum mechanics without Bell inequalities, *Phys. Rev. A* **56**(1), 176–181 (1997).

[7] M. Renninger, Messungen ohne störung des meß objekts, *Zeit. Phys.* **158**, 417–421, (1960).

[8] A. C. Elitzur and L. Vaidman, Quantum mechanical interaction-free measurements *Found. Phys.* **23**(7), 987–997 (1993).

[9] J. A. Wheeler, in: *Mathematical Foundations of Quantum Mechanics*, edited by A. R. Marlov (Academic Press, New York, 1978), pp. 9–48.

[10] J. A. Wheeler, in: *Some Strangeness in the Proportion: A Centennial Symposium to Celebrate the Achievements of Albert Einstein*, edited by H. Woolf (Addison Wesley, Reading, MA, 1980), pp. 341–375.

[11] N. Bohr, *Discussion with Einstein on Epistemological Problems of Atomic Physics*, in: *Albert Einstein: Philosopher-Scientist*, edited by P. A. Schilpp (Harper and Row, New York, 1959), pp. 199–241.

[12] B.-G. Englert, M. O. Scully and H. Walther, The duality in matter and light, *Sci. American*, **271**(6), 56–61 (1994).

[13] M. O. Scully, B.-G. Englert & H. Walther, Quantum optical tests of complementarity, *Nature* **351**, 111–116(1991).

[14] U. Mohrhoff, Restoration of interference and the fallacy of delayed choice: concerning an experiment proposed by Englert, Scully and Walther, *Am. J. Phys.* **64**(12), 1468–1475 (1996).

[15] S. Wiesner, Conjugate coding, *SIGACT News* **15**(5), 78–88 (1983).

[16] D. Deutsch, Quantum communications moves into the unknown, *Phys. World* **6**(6), 22–23 (1993).

[17] C. H. Bennett, G. Brassard, S. Breidbart and S. Wiesner, in: *Advances in Cryptography: Proceedings of Crypto 82* (Plenum, New York, 1983), 267–275.

[18] A. K. Ekert, B. Huttner, G. M. Palma and A. Peres, Eavesdropping on quantum-cryptographical systems, *Phys. Rev. A* **50**(2), 1047–1056 (1994).

[19] C. H. Bennett and G. Brassard, Quantum public key distribution system, *IBM Tech. Disclosure Bull.* **28**(7), 3153–3163 (1985).

[20] G. Brassard, C. Crépeau, R. Jozsa and D. Langlois, in: *Proceedings of the 34th Annual IEEE Symposium on Foundations of Computer Science* (IEEE, Los Alamitos, 1993), pp. 362–371.

[21] A. Yao, in: *Proceedings of the 26th Symposium on the Theory of Computing* (ACM, New York, 1995), pp. 67–75.

[22] H-K. Lo and H. F. Chau, Is quantum bit commitment really possible? *Phys. Rev. Lett.* **78**(17), 3410–3413 (1997).

[23] D. Mayers, Unconditionally secure quantum bit commitment is impossible, *Phys. Rev. Lett.* **78**(17), 3414–3417 (1997).

[24] C. H. Bennett, G. Brassard, R. Jozsa, A. Peres, W. Wooters, Teleporting an unknown quantum state via dual classical and Einstein-Podolski-Rosen channels, *Phys. Rev. Lett.* **70**(13), 1895–1899 (1993).

[25] D. Bouwmeester, J. W. Pan, K. Mattle, M. Eibl, U. Weinfurt, A. Zeilinger, Experimental quantum teleportation, *Nature* **390**, 575–579 (1997).

[26] D. Boschi, S. Branca,, F. De Martini, L. Hardy, S. Popescu, Experimental realization of teleporting an unknown pure quantum state via dual classical and EPR channels, *Phys. Rev. Lett.* **80**(6), 1121–1125 (1998).

[27] G. P. Collins, Quantum teleportation channels opened in Rome and Innsbruck, *Phys. Today* **51**(2), 18–21 (1998).

[28] A. Aspect, P. Grangier and G. Roger, Experimental tests of realistic local theories via Bell theorem, *Phys. Rev. Lett.* **47** (7), 460–463 (1981).

[29] A. Aspect, I. Dalibard and G. Roger, Experimental test of Bell inequalities using time-varying analysis, *Phys. Rev. Lett.* **49** (25) 1804–1807 (1982).

[30] W. Tittel, J. Brendel, B. Gisin, T. Herzog, H. Zbinden, and N. Gisin, Experimental demonstration of quantum correlations over more than 10 kilometers, *Phys. Rev A* **57**(5), 3229–3232 (1998)

[31] S. L. Braunstein, A. Mann and M. Revzen, Maximum violation of Bell's inequalities for mixed states, *Phys. Rev. Lett.* **68**(22), 3259 (1992).

16
WHERE ARE WE NOW?

In this book we discussed only some of the main interpretations of quantum theory and its nonlocality, which manifests itself in situations of the Einstein-Podolsky-Rosen type.[1]

We did not discuss here other situations where nonlocality is present in quantum physics as, e.g., nonlocality due to *nontrivial topology*. An example of this kind of nonlocality is given by the Aharanov-Bohm effect [3] where it is possible to see the influence of a magnetic field on the interference pattern resulting from particles that follow paths, located at points where the magnetic field is zero (while the vector potential is not zero). Another example is the Casimir effect where the boundary conditions (or other topology — torus, etc.) lead to the appearance of non zero stress-energy tensor at points far from the boundaries [4].

Nor we discuss the so-called *nonlocality in time* that appears in some problems of *quantum cosmology* and which is somehow connected with the so-called *time machines* of general relativity[5].

It is also worth mentioning an interesting idea coming from Leinaas[6-8] and Leinaas and Myrheim [9]. According to these authors, the principle of the identity of particles can be understood as the necessity for a *nontrivial topology* in the configuration space of a n-particles system. Even in classical statistical mechanics, in order to be free from Gibbs paradox, we cannot use $\mathcal{C}_{3n}$ as the configuration space for the n-particle system, which is the usual Cartesian product of n one-particle configuration spaces. Instead, we must first identify points in $\mathcal{C}_{3n}$, which are related by permutation of the particle indices, and then use as configuration space, the quotient space $\mathcal{C}_{3n}/S^n$, where S^n is a discrete symmetry group. Quantum theory in $\mathcal{C}_{3n}/S^n$, which has a different topology in relation to $\mathcal{C}_{3n}$, is such that we automatically come to the necessity of having either symmetric (*bosons*) or antisymmetric wave functions (*fermions*) when the particles can move in the 3-dimensional physical space. If the particles can move only in a space of dimension less then 3, then the possibility of *anyons* excitations exists[6–9].

It is necessary to observe here that the configuration space in quantum theory, different from the case of classical statistical mechanics, has a different and more profound meaning. In classical statistical mechanics, it is always possible to describe an n-particle system as moving in a usual 3-dimensional space, and the $3n$-configuration space is just a mathematical *description*. However, in quantum theory, nobody solved successfully the problem of how to understand the n-particle wave function, which "propagates" in the $3n$-dimensional configuration space, as some waves propagating in 3-dimensional ordinary space (see Barut's references in Chapter 10).

This is the reason, that contrary to the cases of the Bohm-Aharanov or Casimir effects, where the region of physical space of interest for the physical phenomena under study has a nontrivial topology, in $3n$-dimensional configuration space, a nontrivial topology can lead to new nonlocality effects, manifested in EPR like situations. So, it is possible that the development of a new science — *quantum topology* — could lead to some new insights in the nonlocality problem of quantum physics [10,11].

As to different interpretations of quantum physics discussed in the book, we arrive at the following conclusions.

1. The *Copenhagen interpretation* in any of its forms is the best interpretation, consistent with the *mathematical formalism* of quantum theory and all the available *experimental evidence* supporting this formalism.

Nevertheless in its form, when only *classical* objects have *existence* status, it leaves the *Schrödinger's cat paradox* unsolved, and as has been discussed in detail in this book, is equivalent to the claim that there are no true *classical* objects if we do not change the mathematical formalism of the quantum theory in some way when dealing with macroscopic bodies. To the authors, more self-consistent seems the *subjective interpretation*, which in whatever of its forms— (a) the *ultimate Ego* of von Neumann, (b) the difference between the *Boolean logic* of human consciousness and the non-Boolean logic of the world, (c) the *objective existence* of many copies of the observer and *subjective* identification of the observer with only one of them —consciousness, or spirit, play an important and fundamental role.

This interpretation surely contradicts philosophical prejudices of the majority of physicists, conditioned by the social life of all western countries, which almost invariably leads people to practice naive materialism as the most popular philosophy. But, philosophical prejudices must be avoided if we intend to understand such an exact science as physics.

Nevertheless, we think that in order for this interpretation to be taken seriously, it will need the support of scientific developments in such sciences as psychology, neurophysiology and psychiatry. While some active scientists in these sciences, like, e.g., Eccles [12] really see new future possibilities in this union of their sciences with quantum physics, other scientists do not have this opinion. Some actively working physicists like Stapp [13], Squires [14] and Penrose [15] also consider this union necessary, although they do not support the subjective interpretation.

Sometimes we hear objections from cosmologists to the Copenhagen interpretation. They insist on the necessity of having the wave function of the universe in quantum cosmology and say that since there is no *external observer* to the universe, it is necessary to invent a new interpretation or new form of quantum formalism. They are then strong supporters of the *many worlds* or the *histories* interpretation. But we can say, that von Neumann's *ultimate Observer*, being *subject* is not *part* of any universe at all, so that it is only a materialistic philosophical prejudice which is working here (a discussion of the role of observer in quantum cosmology can be found in [16]).

Also the *many worlds* (or many points of view) interpretation of quantum theory allows for interesting possibilities. Bell's idea to interpret it as a *many pasts existence* with one *present* may be of interest for cosmology. In a Friedmann universe with the Big Bang as its beginning, there is a property of *causal disconnectedness* of the early universe, as if our present universe really evolved from many, (but not one) early universes [17]. So, there eventually can exist a profound connection between the Copenhagen interpretation in this form and cosmology. One *present* with many *pasts* and many *futures* may be due to the special role of the *subject* who is in the present.

2. We comment also that according to our point of view, as discussed in Chapter 15, the *histories approach* in its actual form, cannot do what their proponents claim. Ingredients outside quantum theory are necessary to give a meaning to the theory, besides many other problems mentioned in Chapter 15.

3. *Realistic interpretations.* All interpretations of this kind, in this or that form, imply in a change of the mathematical formalism of either quantum theory or special relativity or both!

The breakdown of Bell's inequalities shows an ultimate unity between the formalism of quantum mechanics and special relativity. Indeed, examples in de Broglie-Bohm theory, the possibilities of obtaining quantum logical non-distributive structures due to some incompleteness in the Boolean structure of classical objects, the Ghirardi, Rimini, Weber localization processes etc. – all this show that it is special relativity that leads to serious difficulties in relativistic generalizations of *realistic theories*. Comparing these models with existing relativistic quantum theory (especially the elementary particles theory based on the relativistic quantum field theory), we can definitely say: All these models in no sense are on the same level as the standard theory! While giving explanations for some special cases, like *EPR* correlations in, e.g., Aspect's experiment, all these theories in the relativistic domain still leave many problems unsolved, while in standard theory these problems simply do not arise. This statement, of course, does not mean that the standard theory does not have its own problems.

However, as with any new idea, these theories lead to some new predictions, which if confirmed by future experiments, surely will lead to a new fundamental step in our understanding of nature and the human beings taking part in it.

Notes

1. Our text, of course, does not do justice to the many interesting approaches concerning the interpretation of quantum theory that have been proposed in the literature. The interested reader can find many references in [1,2].

References

[1] F. Selleri, *Quantum Paradoxes and Physical Reality*, Kluwer Academic Publishers, Dordrecht, 1990.

[2] A. Jabs, Quantum mechanics in terms of realism, *Phys. Essays* **9**(1), 36–95 (1996).

[3] Y. Aharanov and D. Bohm, Significance of electromagnetic potentials in the quantum theory, *Phys. Rev.* **115**(3) 485–491 (1959).

[4] H. B. G. Casimir, On the attraction between two perfect conducting plates, *Proc. Kon. Nederl. Akad. Wetensc B* **51**, 793–796 (1948).

[5] M. S. Morris and K. S. Thorne, Wormholes, time machines, and the weak energy condition, *Phys. Rev. Lett.* **61**(13), 1446–1449 (1988).

[6] J. Leinaas, Theory of identical particles, *N. Cimento B* **37**(1), 1–23 (1977);

[7] J. Leinaas, Statistics of charge-monopole composite, *N. Cimento A* **47**(1), 19–34 (1978).

[8] J. Leinaas, Topological charges in gauge theories, *Fort. Phys.* **28**(11), 579–631 (1980)

[9] J. Leinaas and R. Myrheim, Quantum theories for identical particles, *Int. J. Mod. Phys. B* **5**(16-17), 2573–2583 (1991).

[10] A. A. Grib and R. R. Zapatrin, Topologimeter and the problem of physical interpretation of topology lattice, *Int. J. Theor. Phys.* **35**(3), 593–604 (1996).

[11] A. A. Grib and R. R. Zapatrin *In Search of Quantum Topology*, RP 72/96, IMECC-UNICAMP (Brazil), 1996.

[12] J. C. Eccles, *How the Self Controls its Brain* (Springer-Verlag, Berlin,1994).

[13] H. P. Stapp, *Mind, Matter, and Quantum Mechanics* (Springer-Verlag, Berlin, 1993).

[14] E. J. Squires, *The Mystery of the Quantum World* (Adam Hilger Ltd., Bristol, 1986).

[15] R. Penrose, *Shadows of the Mind* (Oxford University Press, Oxford,1994).

[16] A. A. Grib, in: *Origins, Time & Complexity, Studies in Science and Theology*, edited by G. V. Coyne, S. J., and K. Schmitz-Moormannn (Labor et Fides, Weil am Rhein, Germany, 1993), pp. 93–113.

[17] A. A. Grib, S. G. Mamayev and V. M. Mostepanenko, *Vacuum Quantum Effects in Strong Fields* (Friedmann Laboratory Publishing, St. Petersburg, 1994).

Appendix A
SET THEORY AND LATTICES

The objective of the Appendices is to present the main definitions and propositions of some of the *mathematical structures* used in the text. With a few exceptions, no attempt was made to present proofs of the propositions, for otherwise the length of the Appendices would be greater than the main text. Any serious student who needs to improve his knowledge about the topics treated below must consult mathematical texts as, e.g., [1-6] and also the following excellent texts on quantum theory[7-10].

A.1 MAIN DEFINITIONS

There are two primitive concepts in set theory, the notion of *set* itself[1] and the notion of *pertinence* as used, e.g., in sentences like "x is an element of the set S."

We write $x \in S$ if x is a member of a given set S and write $x \notin S$ otherwise.

In general a set S is determined by some property P shared by its members and we denote this fact writing $S = \{x | P(x)\}$.

definition 1. Given any property P, the *empty* (or *null*, or *void*) set is the set

$$\emptyset = \{x | P(x) \text{ is false}\}. \tag{A.1}$$

definition 2. Let A and B be two sets such that $\forall x \in A \Rightarrow x \in B$. We say that A is a *subset* of B and denote this fact, writing $A \subseteq B$ or $B \supseteq A$. If it happens that $A \subseteq B$ and $B \supseteq A$, then both sets are equal and we write $A = B$. If $A \subseteq B$ or $A \neq B$ we say that A is a *proper subset* of B and write sometimes $A \subset B$.

The symbol $\subseteq$ (and also $\subset$) is called *set theoretical inclusion*. We observe that given any set S we have $\emptyset \subseteq S$ (or better $\emptyset \subset S$).

definition 3. Let $A \subseteq S$. The mapping

$$\chi_A : A \to S, \tag{A.2}$$
$$\chi_A(x) = 1 \text{ if } x \in A, \chi_A(x) = 0 \text{ if } x \notin A.$$

is called the *characteristic function* of the set A.

definition 4. Given two sets A and B, their *union* is the set denoted $A \cup B$ such that

$$A \cup B = \{x | x \in A \text{ or } x \in B\}. \tag{A.3}$$

definition 5. Given two sets A and B, their *intersection* is the set denoted $A \cap B$ such that

$$A \cap B = \{x | x \in A \text{ and } x \in B\}. \tag{A.4}$$

If $A \cap B = \emptyset$, then A and B are said to be *disjoint*.

definition 6. The *complement* of $A \subseteq S$ is the set denoted A^c (or $S \setminus A$) such that

$$A^c = \{x \notin A, x \in S\}. \tag{A.5}$$

definition 7. Let $A, B \subseteq S$. The *difference* of A and B is the set denoted $A - B$ such that

$$A - B = A \cap B^c = \{x | x \in A \text{ and } x \notin B\}. \tag{A.6}$$

The difference of A and B is also called the *relative complement* of B in A.

definition 8. The *symmetric difference* of A and B ($A, B \subseteq S$) is the set denoted $A \triangle B$ such that

$$A \triangle B = (A - B) \cup (B - A). \tag{A.7}$$

definition 9. Let J be a set whose elements we will call *indexes*. Given a set X, an *indexed family* of elements of X with indexes in J is a mapping $x : J \to X, \alpha \mapsto x_\alpha$. The indexed family is denoted $\{x_\alpha\}_{\alpha \in J}$ or simply $\{x_\alpha\}$ when it is clear who the set J is.

When $J = N = \{1, 2, ...\}$, the indexed family is said to be a *sequence* of elements of X.

definition 10. An indexed family $\{A_\alpha\}$ of subsets of A, with indexes in J is said to be a *partition* of A if $\forall \alpha, \beta \in J, \alpha \neq \beta, A_\alpha \cap A_\beta = \emptyset$ and $\cup_{\alpha \in J} A_\alpha = A$

definition 11. The union and the intersection of the A_α are respectively the sets

$$\begin{aligned}\cup_{\alpha \in J} A_\alpha &= \{x | x \in A_\alpha \text{ for at least one } \in J\}, \\ \cap_{\alpha \in J} A_\alpha &= \{x | x \in A_\alpha \text{ for all } \alpha \in J\}.\end{aligned} \tag{A.8}$$

definition 12. Given a set S, the class of all its subsets is said to be the *power set* of S and is denoted by 2^S.

definition 13. The *Cartesian product* of the n sets $A_1, A_2, ... A_n$ is the set

$$A_1 \times A_2 \times .. \times .A_n = \{x_1, x_2, ..., x_n) | x_i \in A_i, i = 1, 2, ...n\}. \tag{A.9}$$

When $A_i = A, i = 1, 2, ...n$ we write the Cartesian product as A^n.

A.2 POSETS

definition 14. A (binary) *relation* in a set S is a specified subset $R \subseteq S \times S$. If $(x, y) \in R$ we say that x stands in relation with y and write xRy.

definition 15. Let $R \subseteq S \times S$ be a relation in S. It is said to be:
(i) *reflexive* if $xRx, \forall x \in S$,
(ii) *symmetric* if $xRy \Rightarrow yRx$,
(iii) *transitive* if $xRy, yRz \Rightarrow xRz$,
(iv) *antisymmetric* if xRy and $yRx \Rightarrow x = y$.

definition 16. An *equivalence* relation in a set S is a relation in S which is reflexive, symmetric and transitive.

We often use the symbol $x \sim y$ to denote that $x, y \in S$ are equivalent.

proposition 17. An equivalence relation in S leads to a unique partition of S. Conversely, any given partition of S defines an equivalence relation in S.

definition 18. Let $\sim$ be an equivalence relation in S. Let $x \in S$ be an arbitrary but fixed element. Consider the collection of all y such that $x \sim y$. This collection denoted $[x] \subseteq S$ is called the equivalence class of x.

We have,

$$[x] = \{y \in S \text{ and } y \sim x\}. \tag{A.10}$$

definition 19. Let $E \subseteq S \times S$ be an equivalence relation in S. The *quotient set* of S modulo E, denoted by S/E is the class of all distinct equivalence classes induced in S by E. We have,

$$S/E = \{[x] | \, x \in S\}. \tag{A.11}$$

definition 20. The mapping $\pi : S \to S/E; x \mapsto [x]$ is called the canonical mapping (or projection) of S onto S/E.

definition 21. A relation in a set S is called an *order* relation if it is reflexive, antisymmetric and transitive.

The symbol $\leq$ is used to denote in what follows the order relation. If $x, y, z \in S$ we have,

$$\begin{aligned}\text{(i) } & x \leq x, \forall x \in S, \\ \text{(ii) } & x \leq y \text{ and } y \leq x \Rightarrow x = y, \\ \text{(iii) } & x \leq y \text{ and } y \leq z \Rightarrow x \leq z. \end{aligned} \tag{A.12}$$

Sometimes instead of $x \leq y$ (which reads x is less than or equal to y, or x is contained in y) we write $y \geq x$. If $x \leq y$ and $x \neq y$ we write $x < y$. When $x < y$ and there is no z such that $x < z < y$ we say that y *covers* x and write $x \prec y$.

definition 22. Given a set S and an order relation $\leq$ in S, the pair $(S, \leq)$ is called a *poset* (a short form for partially ordered set).[2]

definition 23. Let $(S, \leq)$ be a poset. If there exists $\mathbf{0} \in S$ such that $\mathbf{0} \leq x, \forall x \in S$, then $\mathbf{0}$ is said to be the *first* (or *smallest*, or *least*) element of S. If there exists $\mathbf{1} \in S$ such that $x \leq \mathbf{1}, \forall x \in S$, then $\mathbf{1}$ is said to be the *last* (or *largest*, or *greatest*) element of S.

proposition 24. The first element of S when it exists is unique. The last element of S if it exists, is unique.

definition 25. Let $(S, \leq)$ be a poset. $m \in S$ is called a *minimal* element of S if there is no element in S which is strictly smaller than m, i.e., if $x \leq m$, then $x = m$. An element $m \in S$ is said to be the *maximal* element of S if there is no element of S that is strictly greater than m, i.e., if $m \leq x$, then $m = x$.

definition 26. Let $(S, \leq)$ be a poset and $A \subseteq S$ (eventually $A = S$). $l \in S$ is said to be a *lower bound* of of A if and only if $\forall a \in A \Rightarrow l \leq a$. An element $u \in S$ is called an *upper bound* of A if and only if $\forall a \in A \Rightarrow u \geq a$.

It is important to have in mind that $A \subseteq S$ need not have necessarily any lower bound and even if it does, let us say b, in general $b \notin A$. In this way it is clear that A may have many different lower bounds. An analogous comment is valid regarding upper bounds.

definition 27. Let $(S, \leq)$ be a poset and $A \subseteq S$. Denote by A_L the set of all lower bounds of A. If A_L has a last (greatest) element, then it is called the *infimum* (or the *greatest lower bound*) and is denoted $\inf(A)$. Denote the set of all upper bounds of A by A_U. If A_U has a first (smaller) element then it is called the *supremum* (or *least upper bound*)) denoted $\sup(A)$.

The notations $\vee_{a \in A} a$ and $\wedge_{a \in A} a$ are often used respectively for $\inf(A)$ and $\sup(A)$. When A is a finite countable (infinite countable) set we write for the infimum $\vee_{i=1}^{n} a$ ($\vee_{i=1}^{\infty} a$). For the supremum we write $\wedge_{i=1}^{n} a$ ($\wedge_{i=1}^{\infty} a$).

proposition 29. Let $(S, \leq)$ be a poset and $x, y, z \in S$. Then,

$$\begin{aligned} \text{(i) } \inf\{x, \{y, z\}\} &= \inf\{x, y, z\} \\ \text{(ii) } \sup\{x, \{y, z\}\} &= \sup\{x, y, z\} \end{aligned}$$

A.3 LATTICES

definition 30. A poset $\mathcal{L}$ is called a *lattice* if:
(i) $\exists\, 0 \in \mathcal{L}, \exists\, 1 \in \mathcal{L}$ and $0 \neq 1$,
(ii) each pair $x, y \in \mathcal{L}$, $x \neq y$, has a supremum (also called *union*, *join* or *disjunction*) and denoted by $x \vee y$ and an infimum (also called *intersection*, *meet* or *conjunction*) denoted by $x \wedge y$.

definition 31. If every finite or infinite countable subset $\mathcal{A} \subseteq \mathcal{L}$ has a join and a meet, then $\mathcal{L}$ is said to be a σ-*complete lattice*. If for any nonempty set $\mathcal{A} \subseteq \mathcal{L}$, $\vee_{a \in \mathcal{A}} a$ and $\wedge_{a \in \mathcal{A}} a$ exist, then the lattice is said to be *complete*. If $\mathcal{L}$ is finite countable, then it is said to be *finite*.

definition 32. A lattice $\mathcal{L}$ is said to be *atomic* (or atomistic) if $\forall x \in \mathcal{L} \Rightarrow x \geq a$, where $0 \neq a \in \mathcal{L}$ is an *atom*, i.e., $0 \prec a$.

proposition 33. Let $\mathcal{L}$ be a lattice and $x, y, z \in \mathcal{L}$. The following properties are true:
(i) *idempotency*: $x \vee x = x$, $x \wedge x = x$,
(ii) *commutativity*: $x \vee y = y \vee x$, $x \wedge y = y \wedge x$,
(iii) *associativity*: $x \wedge (y \wedge z) = (x \wedge y) \wedge z$, $x \vee (y \vee z) = (x \vee y) \vee z$,
(iv) *convolution or absorptivity*: $x \wedge (x \vee y) = x$, $x \vee (x \wedge y) = x$,
(v) the statements: (a) $x \leq y$, (b) $x \vee y = y$, (c) $x \wedge y = x$ are equivalent,

(vi) $x \leq y \Rightarrow x \vee z \leq y \vee z$ and $x \wedge z \leq y \wedge z$.

definition 34. A lattice $\mathcal{L}$ is said to be *complemented* if $\forall x \in \mathcal{L}$ has at least one complement, i.e., $\exists\, y \in \mathcal{L} | x \wedge y = \mathbf{0}$ and $x \vee y = \mathbf{1}$. $\mathcal{L}$ is said to be *uniquely complemented* if $\forall x \in L$ has only *one* complement. In what follows we denote a complement of $x \in \mathcal{L}$ as x' ($x' \in \mathcal{L}$).

proposition 35. Let $\mathcal{L}$ be a lattice and $x, y, z \in \mathcal{L}$. Then,

$$\begin{aligned}\text{(i)}\ x \vee (y \wedge z) &\leq (x \vee y) \wedge (x \vee z), \\ \text{(ii)}\ x \wedge (y \vee z) &\geq (x \wedge y) \vee (x \wedge z).\end{aligned} \quad (A.13)$$

proposition 36. Let $\mathcal{L}$ be a lattice and $x, y, z \in \mathcal{L}$ and $x \leq z$. Then,

$$x \vee (y \wedge z) \leq (x \vee y) \wedge z. \quad (A.14)$$

definition 37. A *distributive* lattice $\mathcal{L}$ is a lattice such that $\forall x, y, z \in \mathcal{L}$, the distributive laws hold, i.e.,

$$\begin{aligned}\text{(i)}\ x \vee (y \wedge z) &= (x \vee y) \wedge (x \vee z), \\ \text{(ii)}\ x \wedge (y \vee z) &= (x \wedge y) \vee (x \wedge z)\end{aligned} \quad (A.15)$$

definition 38. A lattice $\mathcal{L}$ is said to be *modular* if $\forall x, y, z \in \mathcal{L}$ and $x \leq z$, we have

$$x \vee (y \wedge z) = (x \vee y) \wedge z. \quad (A.16)$$

proposition 39. If $\mathcal{L}$ is distributive it is also modular.

definition 40. A complemented distributive lattice $\mathcal{B}$ is called a *Boolean algebra*. If $\mathcal{B}$ is a σ−lattice it is called a Boolean σ-algebra.

proposition 41. Every element in a Boolean algebra (or σ-algebra) $\mathcal{B}$ has only *one* complement. Also, if $x, y, z \in \mathcal{B}$ and $x \vee y = x \vee z, x \wedge y = x \wedge z \Rightarrow y = z$.

proposition 42. $\forall x, y \in \mathcal{B}$ (a Boolean lattice) the following properties hold:
(i) $\mathbf{0}' = \mathbf{1}$; (ii) $\mathbf{1}' = \mathbf{0}$; (iii) $(x')' = x$; (iv) $x = y \Leftrightarrow x' = y'$;
(v) $(x \vee y)' = x' \wedge y'$; (vi) $(x \wedge y)' = x' \vee y'$; (vii) $x \leq y \Rightarrow y' \leq x'$;
(viii) the statements: (a) $x \leq y$, (b) $x \wedge y' = \mathbf{0}$, (c) $x' \vee y = \mathbf{1}$, are equivalent.

definition 43. Let $\mathcal{L}_1$ and $\mathcal{L}_2$ be two lattices. A surjective (onto) mapping is called an *epimorphism* (or homomorphism) of $\mathcal{L}_1$ into $\mathcal{L}_2$ if for any at most countable sequence $\{x_i\}, i = 1, 2\ldots$ of $\mathcal{L}_1$,

$$f(\vee_i x_i) = \vee_i f(x_i) \text{ and } f(\wedge_i x_i) = \wedge_i f(x_i) \quad (A.17)$$

If f is an injective (one-to-one) mapping, it is called a *monomorphism*. If f is bijective, it is called an *isomorphism*.

Since $\forall x \in \mathcal{L}_1, x \vee \mathbf{0} = x, x \wedge \mathbf{0} = \mathbf{0}, x \vee \mathbf{1} = \mathbf{1}, x \wedge \mathbf{1} - x$ it follows that for any f as in definition 43,

$$f(\mathbf{0}) = \mathbf{0},\ f(\mathbf{1}) = \mathbf{1} \quad (A.18)$$

definition 44. An isomorphism $d : \mathcal{L}_1 \to \mathcal{L}_2$ is called *dual* if $\forall x, y \in \mathcal{L}_1$ with $x \leq y$ then $d(y) \leq d(x)$.

definition 45. An isomorphism $\mathcal{L} \to \mathcal{L}$ is called an *automorphism*. A dual automorphism $\mathfrak{a}: \mathcal{L} \to \mathcal{L}$ such that $\forall x \in \mathcal{L}$, $\mathfrak{a}(\mathfrak{a}(x)) = x$ is called *involutive*.

proposition 46. If $\mathfrak{a}: \mathcal{L} \to \mathcal{L}$ is an involutive dual automorphism then,

$$\text{(i)} \ \mathfrak{a}(x \vee y) = \mathfrak{a}(x) \wedge \mathfrak{a}(y);$$
$$\text{(ii)} \ \mathfrak{a}(x \wedge y) = \mathfrak{a}(x) \vee \mathfrak{a}(y) \tag{A.19}$$

and the statements

$$\text{(iii)} \ x \leq \mathfrak{a}(x) \Rightarrow x = 0,$$
$$\text{(iv)} \ \forall x \in \mathcal{L} \Rightarrow x \wedge \mathfrak{a}(x) = 0, \tag{A.20}$$
$$\text{(v)} \ \forall x \in \mathcal{L} \Rightarrow x \vee \mathfrak{a}(x) = 1$$

are equivalent.

definition 47. A lattice $\mathcal{L}$ together with an involute dual automorphism $\mathcal{L} \to \mathcal{L}$ is said to be *orthocomplemented*. The involutive dual automorphism is called an *orthocomplementation* and is denoted by $\perp: \mathcal{L} \ni x \mapsto x^\perp \in \mathcal{L}$. $x^\perp$ is said to be the orthocomplement of x. In what follows, an orthocomplemented lattice will be denoted by $\mathcal{L}_o$.

Since $x^\perp$ is a complement (definition 34) we denote in the text $x^\perp$ by x' when no confusion arises.

definition 48. We say that $x, y \in \mathcal{L}_o$ are *orthogonal* (or *disjoint*) and write $x \perp y$ if $x \leq y^\perp$.

It follows immediately from definition 48 that $x \perp y \Rightarrow y \perp x$.

definition 49. A *weakly modular* lattice denoted $\mathcal{L}_w$ in what follows is an orthocomplemented lattice, such that $\forall x, y \in \mathcal{L}_w$ then

$$x \leq y \Rightarrow y = x \vee (x' \wedge y). \tag{A.21}$$

definition 50. A *quasi-modular* lattice is an orthocomplemented lattice, such that $\forall x, y, z \in \mathcal{L}$ then

$$x \leq y \leq z' \Rightarrow x = (x \vee z) \wedge y. \tag{A.22}$$

definition 51. An *orthomodular* lattice denoted $\mathcal{L}_{om}$ in what follows is an orthocomplemented lattice such that $\forall y, z \in L$ then $\forall x \leq z$ we have $y \perp z \implies x \vee (y \wedge z) = (x \vee y) \wedge z$.

proposition 52. An orthocomplemented lattice $\mathcal{L}$ is weakly modular if and only if $\forall x, y \in \mathcal{L}$ then

$$x \leq y \Rightarrow x = y \wedge (x \vee y'). \tag{A.23}$$

proposition 53. For an orthocomplement lattice $\mathcal{L}$ we have that:

Orthomodularity $\Leftrightarrow$ *quasi-modularity* $\Leftrightarrow$ *weak modularity*.

proposition 54. Every orthocomplemented lattice is orthomodular.

proposition 55. Let $x, y \in \mathcal{L}_o$. Then
$$(x \wedge y) \vee (x \wedge y') \leq x. \tag{A.24}$$

definition 56. Let $x, y \in \mathcal{L}_o$. We say that x is *compatible* with y and write $x \leftrightarrow y$ if
$$(x \wedge y) \vee (x \wedge y') = x. \tag{A.25}$$

definition 57. Let $x, y \in \mathcal{L}_o$. We say that x is *commensurable* with y and write $x \leftrightsquigarrow y$ if
$$x \vee (x \wedge y') = y \vee (y' \wedge x). \tag{A.26}$$

definition 58. Let $x, y \in \mathcal{L}_o$. We say that x is *commeasurable* with y and write $x \sim y$ if $\exists\, x_1, y_1, z \in \mathcal{L}_o$ with $x_1 \perp y_1$, $x_1 \perp z$, $y_1 \perp z$ such that $x = x_1 \vee z$ and $y = y_1 \vee z$.

proposition 59. Let $x, y \in \mathcal{L}_o$. Then, $x \leftrightarrow y \Rightarrow x \leftrightarrow y'$.
proposition 60. Let $x, y \in \mathcal{L}_{om}$. Then, $x \leftrightarrow y \Rightarrow y \leftrightarrow x$ and $x \leq y \Rightarrow x \leftrightarrow y$.
proposition 61. Let $x, y \in \mathcal{L}_o$. If $x \leq y \Rightarrow x \leftrightarrow y$ then $\mathcal{L}_o$ is a $\mathcal{L}_{om}$.
proposition 62. Let $x, y \in \mathcal{L}_o$. If $x \leftrightsquigarrow y \Rightarrow y \leftrightsquigarrow x$.
proposition 63. Each one of lattice relations $\leftrightarrow$ or $\leftrightsquigarrow$ or $\sim$ in $\mathcal{L}_{om}$ are such that it implies the other two.

definition 64. A subset $\mathcal{S} \subseteq \mathcal{L}$ is said to be a *sublattice* of $\mathcal{L}$ if $\mathcal{S}$ is itself a lattice with respect to the lattice operations of $\mathcal{L}$.

proposition 65. Let $\{\mathcal{L}_\alpha\}$ be an indexed family of subsets of $\mathcal{L}$. Then the *set theoretical intersection* $\cap_\alpha \mathcal{L}_\alpha$ is a sublattice of $\mathcal{L}$.

definition 66. Let $S \subseteq \mathcal{L}$ be an arbitrary subset of $\mathcal{L}$. We call the lattice *generated* by S, denoted $spanS$, the set theoretical intersection of all the lattices which contain S.

proposition 67. Let $x, y \in \mathcal{L}_o$. Then $x \sim y$ if and only if the lattice generated by $\{x, x', y, y'\}$ is a Boolean (algebra) lattice.

definition 68. An element $c \in \mathcal{L}_o$ is said to be *central* if $c \sim x, \forall x \in \mathcal{L}_o$. The set
$$\mathbf{C}(\mathcal{L}_o) = \{c \in \mathcal{L} | c \sim x, \forall x \in \mathcal{L}_o\} \tag{A.27}$$
is called the center of $\mathcal{L}_o$. If the center is *trivial*, i.e., $\mathbf{C}(\mathcal{L}_o) = \{\mathbf{0}, \mathbf{1}\}$, then $\mathcal{L}_o$ is said to be *irreducible*. If $\mathbf{C}(\mathcal{L}_o) \neq \{\mathbf{0}, \mathbf{1}\}$, then $\mathcal{L}_o$ is said to be *reducible*.

definition 69. Let $\mathcal{L}_1, \mathcal{L}_2, \ldots \mathcal{L}_n$ be lattices. The Cartesian product $\mathcal{L} = \mathcal{L}_1 \times \mathcal{L}_2 \times \ldots \mathcal{L}_n$ is said to be the *direct union* of the $\mathcal{L}_i$ if it is equipped with the order relation
$$(x_1, x_2, \ldots x_n) \leq (y_1, y_2, \ldots y_n) \text{ if } x_i \leq x_{i+1}, y_i \leq y_{i+1}$$
$$\text{with } x_i, y_i \in \mathcal{L}_i,\, i = 1, \ldots n. \tag{A.28}$$

and if the meet and join are given by
$$(x_1, x_2, \ldots, x_n) \wedge (y_1, y_2, \ldots y_n) = (x_1 \wedge y_1, x_2 \wedge y_2, \ldots, x_n \wedge y_n),$$
$$(x_1, x_2, \ldots, x_n) \vee (y_1, y_2, \ldots, y_n) = (x_1 \vee y_1, x_2 \vee y_2, \ldots, x_n \vee y_n). \tag{A.29}$$

proposition 70. If a lattice $\mathcal{L}$ is a direct union, then its center is non trivial, i.e., the lattice is reducible.

A.4 THE BOOLEAN ALGEBRA OF SETS

proposition 71. Let $A, B, C \in 2^S$. The following properties hold true:

$$\begin{aligned}
\text{idempotent laws} &: A \cup A = A, A \cap A = A, \\
\text{associative laws} &: (A \cup B) \cup C = A \cup (B \cup C), \\
&\quad (A \cap B) \cap C = A \cap (B \cap C), \\
\text{commutative laws} &: A \cup B = B \cup A, \ A \cap B = B \cap A, \\
\text{distributive laws} &: A \cup (B \cap C) = (A \cup B) \cap (A \cup C), \\
&\quad A \cap (B \cup C) = (A \cap B) \cup (A \cap C), \\
\text{identity laws} &: A \cup \emptyset = A, \ A \cup S = S, \ A \cap \emptyset = \emptyset, \ A \cap S = A, \\
\text{complement laws} &: A \cup A^c = S, \ (A^c)^c = A, \ A \cap A^c = \emptyset, \\
&\quad S^c = \emptyset, \ \emptyset^c = S \\
\text{De Morgan laws} &: (A \cup B)^c = A^c \cap B^c, (A \cap B)^c = A^c \cup B^c
\end{aligned} \quad (A.30)$$

proposition 72. Any class $\mathbf{C}$ of subsets of X equipped with the operations $\cup$ and $\cap$ such that

$$\begin{aligned}
A, B &\in \mathbf{C} \Rightarrow A \cup B \in \mathbf{C} \\
A, B &\in \mathbf{C} \Rightarrow A \cap B \in \mathbf{C} \\
A &\in \mathbf{C} \Rightarrow A^c \in \mathbf{C}
\end{aligned} \quad (A.31)$$

is a Boolean algebra.

The proof of proposition 72 is trivial; we only need to recall the definition of Boolean algebra given above and identify the set operations $\cup$, $\cap$, $\subseteq$ and the set complementation c with the lattice operations $\vee, \wedge, \leq$ and $'$. Also we identify $\emptyset$ with $\mathbf{0}$, $\mathbf{1}$ with S.

A.5 BOREL SETS, MEASURES AND INTEGRATION

A basic fact about Boolean algebras is that these structures can be identified with a certain class of rings, which enables us to study Boolean algebras using the well developed theory of rings.

definition 73. Let $\mathfrak{R}$ be a non empty class of subsets of a given set S. $\mathfrak{R}$ is said to be a ring of sets if the following conditions hold:

$$\begin{aligned}
\forall A, B \in \mathfrak{R} &\Rightarrow A \Delta B \in \mathfrak{R}, \\
\forall A, B \in \mathfrak{R} &\Rightarrow A \cap B \in \mathfrak{R}.
\end{aligned} \quad (A.32)$$

From definition 63, it is easy to verify that for any finite collection of subsets $A_i, i = 1, 2, ...n$ we have

$$\cup_{i=1}^n A_i \in \mathfrak{R}, \ \cap_{i=1}^n A_i \in \mathfrak{R}. \quad (A.33)$$

definition 74 A ring of sets $\mathfrak{R}$ is a Boolean ring if it is a ring with identity, where the ring operations $+$ (sum) and $\cdot$ (product, denoted by juxtaposition) are such that

$$A + B = A \Delta B$$
$$AB = A \cap B \qquad (A.34)$$

and such that $\forall A \in \mathfrak{R} \Rightarrow A^2 = A$

We can easily show that starting with a Boolean ring $\mathfrak{R}$ and using the definitions $A \wedge B = AB$ and $A \vee B = A + B + AB$ we can convert it into a Boolean algebra.

definition 75. If for every countable sequence $\{A_i\}, i = 1, 2, ..., A_i \in \mathfrak{R}$ it follows that

$$\cup_{i=1}^{\infty} A_i \in \mathfrak{R}, \ \cap_{i=1}^{\infty} A_i \in \mathfrak{R} \qquad (A.35)$$

then $\mathfrak{R}$ is said to be a σ-ring.

proposition 76. If a σ-ring $\mathfrak{R}$ is such that $S \in \mathfrak{R}$, it has a structure of a Boolean σ-algebra.

proposition 77. Let **C** be any class of sets and let $\{R_\alpha\}$ be an indexed family of all rings containing **C**. Then $\mathfrak{R}(\mathbf{C}) = \cap_\alpha R_\alpha$ is also a ring called the ring *generated* by **C**. The ring generated by **C** is the smallest ring containing the class **C** and it is unique.

definition 78. Let X be a set. A *Borel field* $\mathfrak{B}$ on X is a class of subsets of X called *Borel sets* such that the following axioms hold:

(i) If $B_i \in \mathfrak{B}, i = 1, 2, ...$, then all finite union $\cup_{i=1}^{n} B_i$ and even the countable union $\cup_{i=1}^{\infty} B_i$ belong to $\mathfrak{B}$.

(ii) If $B \in \mathfrak{B}$ then $B' \in \mathfrak{B}$.

definition 79. A pair $(X, \mathfrak{B})$ is called a *Borel space*.

As an important example of Borel sets, consider the real line $\mathcal{R} = \{x| -\infty < x < \infty\}$ and let **C** be the class of all bounded semi-closed intervals

$$|a, b) = \{x| - a \leq x < b\} \qquad (A.36)$$

We can verify that the class generated by the σ-ring $\mathfrak{R}(\mathbf{C})$ is a Borel field on $\mathcal{R}$. Then $\{\mathcal{R}, \mathfrak{R}(\mathbf{C})\}$ is said to be the Borel space of the reals. It is important to have in mind that there are many other kinds of Borel sets of the reals[3].

definition 80. A *measurable space* is a pair $(X, \mathfrak{M})$ where $\mathfrak{M}$ is a σ-ring on the set X with the property that X is the union (not necessarily countable) of sets taken from the class $2^{\mathfrak{M}}$. The elements of $\mathfrak{M}$ are said to be the *measurable sets* of X.

definition 81. Let $(X, \mathfrak{B}_X)$ and $(Y, \mathfrak{B}_Y)$ be two Borel spaces. A mapping

$$f . (X, \mathfrak{B}_X) \to (Y, \mathfrak{B}_Y) \qquad (A.37)$$

is said to be a Borel function if and only if for $\forall B_Y \in \mathfrak{B}_Y$ we have that $f^{-1}(B_Y) \in \mathfrak{B}_X$. We often write eq.(A.37) as $f : \mathfrak{B}_X \to \mathfrak{B}_Y$.

definition 82. Let $(X, \mathfrak{M})$ be a measurable space. A *measure* on $\mathfrak{M}$ is a mapping

$$\mu : \mathfrak{M} \to \mathcal{R} \qquad (A.38)$$

satisfying the following properties:

(i) $\forall B \in \mathfrak{M}, \mu(B) \geq 0$,

(ii) $\mu(\emptyset) = 0$,

(iii) If $B_i \in \mathfrak{M}$, $i = 1, 2...$ and $B_i \cap B_j = \emptyset$ for $i \neq j$ then $\mu(\cup_{i=1}^{\infty} B_i) = \sum_{i=1}^{\infty} \mu(B_i)$.

Since the codomain of μ is $\mathcal{R}$ we may have sets whose measure is $+\infty$. Also it may happen that $\sum_{i=1}^{\infty} \mu(B_i)$ is infinite.

definition 83. A set $B \in \mathfrak{M}$ is said to be of *zero measure* or a *null set* with respect to μ if $\mu(B) = 0$.

Also a set $B \in \mathfrak{M}$ is said to have *finite measure* if $\mu(B) \neq \infty$.

In what follows unless explicitly stated we suppose that all measures are finite, i.e., $\forall B \in \mathfrak{M}$, $\mu(B) \neq \infty$.

definition 84. A triple $(X, \mathfrak{M}, \mu)$ where $(X, \mathfrak{M})$ is a measurable space and μ is a measure on $\mathfrak{M}$, is said to be a *measure space*.

definition 85. Take $X = \mathcal{R}$ (the real line). A *Lebesgue measure* μ is a measure where $\mathfrak{M}$ is the σ-ring $\mathfrak{R}(\mathbf{C})$ whose elements are the Borel sets of the real line and

$$\mu\{|a,b)\} = b - a \tag{A.39}$$

definition 86. Take $X = \mathcal{R}$ (the real line). A *Lebesgue-Stieltjes measure* μ is a measure where $\mathfrak{M}$ is the σ-ring $\mathfrak{R}(\mathbf{C})$ whose elements are the Borel sets of the real line and

$$\mu\{|a,b)\} = \varphi(b) - \varphi(a) \tag{A.40}$$

where $\varphi : \mathcal{R} \to \mathcal{R}$ is such that $\lim_{\varepsilon \to 0} \varphi(\lambda + \varepsilon) = \varphi(\lambda)$.

Another important measure used in this book is the *probability measure* on the lattice of propositions of a quantum mechanical system whose possible states are elements of a Hilbert space $\mathcal{H}$ which we describe in Chapter 9. Also the concept of measure space permits us to introduce the Lebesgue and Lebesgue-Stieltjes integrals of functions which are more general than the Riemann integral of elementary calculus and which are *crucial* for the definition of the function realization of a Hilbert space and other function spaces that are fundamental ingredients of quantum theory.

definition 87. Let $(X, \mathfrak{M}, \mu)$ be a measurable space. A measurable function on X is a mapping

$$f : X \to \mathcal{R} \tag{A.41}$$

such that for every Borel set $B \in \mathfrak{R}(\mathbf{C})$ (on the real line), the set $f^{-1}(B) \in \mathfrak{M}$ is measurable.

We see at once that the characteristic function χ_A (see definition 3) of a set $A \in \mathfrak{M}$ is measurable since for every Borel set $B \in \mathfrak{R}(\mathbf{C})$ we have

$$\chi_A^{-1}(B) = \begin{cases} A & \text{if } x \in A \\ \emptyset & \text{if } x \notin A \end{cases} \tag{A.42}$$

Let $(X, \mathfrak{M}, \mu)$ be a measurable space and let $f, g : X \to \mathcal{R}$ be two arbitrary measurable functions. We verify at once that the functions $f + g$ and fg are measurable.

definition 88. Let $(X, \mathfrak{M}, \mu)$ be a measurable space. An *integrable simple function* is a mapping $f : X \to \mathcal{R}$ such that

$$f(x) = \sum_{i=1}^{n} a_i \chi_{A_i}(x), \ a_i \in \mathcal{R}, \ A_i \in \mathfrak{M} \text{ and } A_i \cap A_j = \emptyset, i \neq j \qquad (A.43)$$

It is trivial to verify that any simple function is measurable.

definition 89. Let $(X, \mathfrak{M}, \mu)$ be a measurable space and let $f : X \to \mathcal{R}$ any integrable single function like in eq.(A.43). By definition the integral of f is

$$\int_X f d\mu = \sum_{i=1}^{n} a_i \mu(A_i) < \infty. \qquad (A.44)$$

definition 90. Let $(X, \mathfrak{M}, \mu)$ be a measurable space and let $f : X \to \mathcal{R}$ be a measurable function. We say that a sequence $\{f_i\}$, $i = 1, 2...$, of measurable finite simple functions on X is a *Cauchy sequence in measure* (or converge in the measure) to the measurable function f if for $\forall x \in X$ if $\forall \varepsilon > 0$

$$\lim_{i \to \infty} \mu(\{x : |f(x) - f_i(x)| \geq \varepsilon\}) = 0. \qquad (A.45)$$

proposition 91. Let $(X, \mathfrak{M}, \mu)$ be a measurable space and let $f : X \to \mathcal{R}$ be a a finite value measurable function. f is said to be integrable if there exists a sequence $\{f_i\}$, $i = 1, 2...$, of integrable simple functions on X which converge in the measure to f and

$$\int_{S \subseteq X} f d\mu = \lim_{i \to \infty} \int_{S \subseteq X} f_i d\mu. \qquad (A.46)$$

We observe that it is easy to verify that the rules for integration of the sum $(f + g)$ and product (fg) of finite value measurable functions are equal to the corresponding ones in Riemann integration. Also,

$$\int_{S \subseteq X} f d\mu \leq \int_{S \subseteq X} |f| d\mu. \qquad (A.47)$$

definition 92. If $X \subseteq \mathcal{R}$, the real line and $\mathfrak{M} = \mathfrak{R}(\mathbf{C})$ and μ is the Lebesgue measure then the integral in eq.(A.46) is called the Lebesgue integral of f.

definition 93. If $X \subseteq \mathcal{R}$, the real line and $\mathfrak{M} = \mathfrak{R}(\mathbf{C})$ and if the Lebesgue-Stieltjes measure is generated by a non decreasing function $\rho(x)$ then eq.(A.45) is denoted as

$$\int_X f d\mu_\rho \text{ or } \int_X f(x) d\mu_\rho(x) \text{ or } \int_X f d\rho(x), \qquad (A.48)$$

and is called the Lebesgue-Stieltjes integral of f.

definition 94. If f is a complex measurable function on $(X, \mathfrak{M}, \mu)$, i.e., the real $(\text{Re} f)$ and imaginary $(\text{Im} f)$ parts of f are measurable functions, then its integral of f is given by

$$\int_{S \subseteq X} f d\mu = \int_{S \subseteq X} \text{Re} f d\mu + i \int_{S \subseteq X} \text{Im} f d\mu. \qquad (A.49)$$

definition 95. Let $(X, \mathfrak{M}, \mu)$ be a measurable space. We call L^1 the function space of all complex integrable functions $f : X \to \mathcal{R}$.

proposition 96. L^1 is a linear manifold, i.e., $\forall f, g \in L^1$ and $c \in \mathcal{C}$ (the complex field)

$$f + g \in L^1 \text{ and } cf \in L^1. \tag{A.50}$$

definition 97. Let $f \in L^1$. The *norm* of f is by definition

$$\| f \| = \int_X |f| d\mu. \tag{A.51}$$

definition 98. Let $f, g \in L^1$. The distance between f and g is $\| f - g \|$, if $\| f - g \| = 0$, f is said to be *equivalent* to g. They differ at most on a set of zero measure.

definition 99. A space function is said to be *complete* if for every sequence $\{f_i\}$ that has a limit, i.e., if $\| f_i - g_j \| \to 0$ when $i, j \to \infty$ there exists an integrable function f such that

$$\lim_{i \to \infty} |f - f_i| = 0. \tag{A.52}$$

definition 100. Let $(X, \mathfrak{M}, \mu)$ be a measurable space. We call $L^2(X, \mathfrak{M}, \mu)$ the function space of all complex integrable functions $f : X \to \mathcal{R}$ which are such that the square of its modulus is integrable. We define the *norm* of $f \in L^2(X, \mathfrak{M}, \mu)$ by

$$\| f \| = \int_X |f|^2 d\mu < \infty. \tag{A.53}$$

When the space $L^2(X, \mathfrak{M}, \mu)$ is equipped with a positive inner product defined for $\forall f, h \in L^2(X, \mathfrak{M}, \mu)$ by

$$(f, h) = \int_X f^* h d\mu, \tag{A.54}$$

satisfying the properties in (i) of definition 1 of Appendix B it becomes a *Hilbert space* as the reader can verify without difficulties.

We end this appendix by calling the reader's attention to the fact that besides the generalization of the Riemann integral presented above, there also exists a generalization of the concept of an indefinite integral. This permits us to introduce the so called *Radon-Nikodym derivative* of an integrable function. We are not going to present the corresponding definitions here since they are not needed in the main text. However, we emphasize that any serious student of quantum theory must know this concept well which together with the other topics of this appendix may be found in many excellent books, some of them quoted in the references.

Notes

1. We call the reader's attention to the fact there are naive definitions of sets like the one given by Cantor: "A set is a collection of definite and distinct objects of our intuition or of our thought", which lead to paradoxes[11]. In particular, an arbitrary collection is not necessarily a set. In the use made of set theory in this book, all sets involved occur in the context of a single (universal) set, from which they are parts. Then we can use the word set as synonymous for aggregate, family, collection. The word class is reserved for a collection of sets.

2. When the meaning of the order relation $\leq$ is clear we simply write S for the poset $(S, \leq)$.

References

[1] G. Birkhoff, *Lattice Theory*, AMS Colloquium Publ. **25**, third edition (AMS, Providence, Rhode Island, 1993).

[2] Y. Choquet-Bruhat, C. DeWitt-Morette and M. Dillard-Bleick, *Analysis, Manifolds and Physics*, revised edition (North Holland Publ. Co. Amsterdam, 1997).

[3] P. Roman, *Some Modern Mathematics for Physicists and other Outsiders*, vols. 1 and 2 (Pergamon Press Inc., New York, 1975).

[4] I. R. Porteous, *Topological Geometry*, second edition (Cambridge Univ. Press, Cambridge, 1981).

[5] G. F. Simmons, *Introduction to Topology and Modern Analysis*, (McGraw-Hill Book Co., Singapore, 1963).

[6] A. E. Taylor, *Introduction to Functional Analysis* (J. Wiley & Sons, New York, 1958).

[7] N. N. Bogolubov, A. A. Logunov and I. T. Todorov, *Introduction to Axiomatic Quantum Field Theory* (W. A. Benjamin Inc., Reading, MA, 1975).

[8] J. M. Jauch, *Foundations of Quantum Theory* (Addison-Wesley Publ. Co., Reading, MA, 1968).

[9] M. Jammer, *The Philosophy of Quantum Mechanics* (J. Wiley & Sons, New York, 1974).

[10] G. W. Mackey, *Mathematical Foundations of Quantum Mechanics* (W. A. Benjamin, New York, 1963).

[11] V. S. Varadarajan, *Geometry of Quantum Theory*, vols. 1 and 2 (D. van Nostrand Co. Inc., Princeton, 1968).

[12] R. Rucker, *Infinity and the Mind* (Bantan Books, New York, 1983).

Appendix B
HILBERT SPACES

References for Appendix B include all the ones of Appendix A and some other specific ones.

B.1 DEFINITION OF HILBERT SPACE

definition 0. A linear (vector) space V over a commutative field $\mathcal{K}$ is a non empty set whose elements are called vectors, denoted by boldface letters like $\mathbf{v}, \mathbf{w}, ...$, etc.,[1] and such that $\forall \mathbf{v}, \mathbf{w}, \mathbf{z} \in V$ and $0, 1, k, m \in \mathcal{K}$ there exists a mapping (called scalar multiplication)[2],

$$\mathcal{K} \times V \to V; \; (k, \mathbf{v}) \mapsto k\mathbf{v} = \mathbf{v}k,$$

such that the following properties hold:

$$
\begin{aligned}
\text{(i)} \; \mathbf{v} + \mathbf{w} &= \mathbf{w} + \mathbf{v}, \\
\text{(ii)} \; (\mathbf{v} + \mathbf{w}) + \mathbf{z} &= \mathbf{v} + (\mathbf{w} + \mathbf{z}), \\
\text{(iii)} \; \exists \, \mathbf{0} &\in V \text{ such that } \mathbf{0} + \mathbf{v} = \mathbf{v}, \\
\text{(iv)} \; \exists \, (-\mathbf{v} &= -1\,\mathbf{v}) \in V \text{ such that } \mathbf{v} + (-\mathbf{v}) = \mathbf{0}, \\
\text{(v)} \; 1\,v &= v, \\
\text{(vi)} \; k(m\mathbf{v}) &= (km)\mathbf{v}, \\
\text{(vii)} \; (k + m)\mathbf{v} &= k\mathbf{v} + m\mathbf{v}, \\
\text{(viii)} \; k(\mathbf{v} + \mathbf{w}) &= k\mathbf{v} + k\mathbf{w}.
\end{aligned}
$$

Note also that $0\mathbf{v} = \mathbf{0}, \forall \mathbf{v} \in V$.

definition 1. A *Hibert space* (denoted $\mathcal{H}$) is a complex linear space (i.e., $\mathcal{K} = \mathcal{C}$) for which the following axioms hold:

(i) There exists in $\mathcal{H}$ a strictly *positive inner product* (sometimes called *scalar product*), i.e., a mapping

$$(,) : \mathcal{H} \times \mathcal{H} \to \mathcal{C}, \qquad (B.1)$$

satisfying the properties

$$(|x\rangle, |y\rangle) = (|y\rangle, |x\rangle)^*, \tag{B.2}$$
$$(|x\rangle, |y\rangle + |z\rangle) = (|x\rangle, |y\rangle) + (|x\rangle, |z\rangle), \tag{B.3}$$
$$(|x\rangle, c|y\rangle) = c(|x\rangle, |y\rangle), \forall c \in C. \tag{B.4}$$

(ii) There exists a *norm* function in $\mathcal{H}$, i.e., a mapping

$$\| \: \|: H \to \mathcal{R}^+ + \{0\}, \tag{B.5}$$

such that

$$\| \: |x\rangle \: \| = \sqrt{(|x\rangle, |x\rangle)}, \tag{B.6}$$

and $\| \: |x\rangle \: \| = 0 \Rightarrow |x\rangle = 0$,

(iii) $\mathcal{H}$ is *separable*, i.e., there exists a sequence $\{|\phi_i\rangle\}.i = 1, 2...., |\phi_i\rangle \in \mathcal{H}$ *dense* in $\mathcal{H}$ in the following sense: given $\forall |\phi\rangle \in \mathcal{H}$ and $\varepsilon > 0$, there exists at least one $|\phi_i\rangle$ in the sequence such that

$$\| \: |\phi_i\rangle - |\phi\rangle \: \| < \varepsilon, \tag{B.7}$$

(iv) $\mathcal{H}$ is *complete*, i.e., any sequence such that

$$\lim_{i,j \to \infty} \| \: |\phi_i\rangle - |\phi_j\rangle \: \| = 0, \tag{B.8}$$

defines a unique $|\phi_i\rangle \in \mathcal{H}$ such that

$$\lim_{i \to \infty} \| \: |\phi_i\rangle - |\phi\rangle \: \| = 0. \tag{B.9}$$

definition 2. When eq.(B.9) holds we write $|\phi_i\rangle \to |\phi\rangle$ and we say that the sequence $\{|\phi_i\rangle\}$ is *strongly* convergent.

We can easily prove the following inequalities,

$$|(|\phi\rangle, |\psi\rangle)| \leq \| \: |\phi\rangle \: \| \| \: |\psi\rangle \: \|, \tag{B.10}$$
$$\| \: (|\phi\rangle + |\psi\rangle) \: \| \leq \| \: |\phi\rangle \: \| + \| \: |\psi\rangle \: \|, \tag{B.11}$$

which are known respectively as the Schwartz and Minkowski inequalities.

definition 3. $|\phi\rangle$ and $|\psi\rangle)$ are said to be *orthogonal* if $(|\phi\rangle, |\psi\rangle) = 0$. We sometimes write $|\phi\rangle \perp |\psi\rangle$.

A sequence $\{|\phi_i\rangle\}, i = 1, 2...., |\phi_i\rangle \in \mathcal{H}$ is said to be orthonormal if

$$(|\phi_i\rangle, |\phi_j\rangle) = \delta_{ij}. \tag{B.12}$$

proposition 4. Given $|\phi\rangle \in \mathcal{H}$ and an orthonormal sequence $\{|\phi_i\rangle\}, i = 1, 2....,$ $|\phi_i\rangle \in \mathcal{H}$ we have

$$\sum_{i=1}^{\infty} |(|\phi_i\rangle, |\phi\rangle)|^2 \leq \| \: |\phi\rangle \: \|^2. \tag{B.13}$$

which is called Bessel's inequality.

definition 5. When the equality sign holds in eq.(B.13), the sequence $\{|\phi_i\rangle\}$ is said to be *complete*.

proposition 6. When the sequence $\{|\phi_i\rangle\}$ is complete the partial sums given by the functions

$$|\varphi_i\rangle = \sum_{i=1}^{\infty}(|\phi_i\rangle,|\phi\rangle)|\phi_i\rangle, \qquad (B.14)$$

converge strongly to $|\phi\rangle$ and we write

$$|\phi\rangle = \sum_{i=1}^{\infty}(|\phi_i\rangle,|\phi\rangle)|\phi_i\rangle. \qquad (B.15)$$

definition 7. In any linear vector space V over a (commutative) field $\mathcal{K}$ a finite or infinite sequence of vectors $\{|\phi_i\rangle\}$, $|\phi_i\rangle \in \mathcal{H}$ is called *linearly independent* if and only if $\sum a_i|\phi_i\rangle = 0 \Rightarrow a_i = 0$, for all i in the index set and $a_i \in \mathcal{K}$. If $\sum a_i|\phi_i\rangle = 0$ and not all $a_i = 0$ then the set $\{|\phi_i\rangle\}$ is said to be *linearly dependent*.

definition 8. The maximal number n of linearly independent vectors in V is called its *dimension*. When n is finite, the space is said to be of *finite* dimension. When $n = \infty$ the space is said to be of *infinite* dimension. Hilbert spaces can be of finite or infinite dimension. When $n < \infty$ the separability and completeness of $\mathcal{H}$ follows from properties (i) and (ii) in definition 1, but when $n = \infty$, this is not the case.

B.2 LINEAR MANIFOLDS AND SUBSPACES

definition 9. Let $M \subseteq \mathcal{H}$ be a subset of $\mathcal{H}$. M is said to be a *linear manifold* if

$$\begin{aligned}&(i)\ \forall |\phi\rangle \in M \Rightarrow c|\phi\rangle \in M,\ \forall c \in \mathcal{C},\\ &(ii)\ \forall |\phi\rangle, |\psi\rangle \in M \Rightarrow |\phi\rangle + |\psi\rangle \in M.\end{aligned} \qquad (B.16)$$

definition 10. Let $S \subseteq \mathcal{H}$ be an arbitrary set of vectors. The smallest linear manifold containing S (which exists and is unique) will be denoted by $\mathcal{M}_S$ or $[S]$ and will be called the linear manifold *spanned* by S.

In an arbitrary Hilbert space $\mathcal{H}$, it may happen that all sequences with only a finite number of non zero components are a linear manifold. However these sequences are not complete, and then the linear manifold is not in general a subspace of $\mathcal{H}$.

definition 11. A *limit vector* of a linear manifold M is a vector $|\phi\rangle \in \mathcal{H}$ such that a sequence $\{|\phi_i\rangle\}$, $i = 1, 2....$, $|\phi_i\rangle \in M$ exists such that $|\phi_i\rangle \to |\phi\rangle$.

definition 12. A *closed* linear manifold (or *subspace*) of $\mathcal{H}$ is a linear manifold M in which every *limit vector* $|\phi\rangle$ of M is such that $|\phi\rangle \in M$. Adding to a linear manifold M all its limit vectors we obtain its *closure* $\mathcal{M} = \overline{M}$ which is the smallest subspace containing M.

definition 13. Given any set $S \subseteq \mathcal{H}$, the set of all vectors orthogonal to S is called the orthogonal complement of S and is denoted

$$S^\perp = \{|\phi\rangle | (|\phi\rangle, |\psi\rangle) = 0, \forall |\psi\rangle \in S\}. \qquad (B.17)$$

We can easily verify that $S^\perp$ is a subspace of $\mathcal{H}$ and directly from eq.(B17) we get

$$\{|0\rangle\}^\perp = \mathcal{H}, \quad \mathcal{H}^\perp = \{|0\rangle\}, \tag{B.18}$$

$$S \cap S^\perp = \{|0\rangle\}, \tag{B.19}$$

$$S_1 \subseteq S_2 \Rightarrow S_2^\perp \subseteq S_1^\perp, \quad \forall S_1, S_2 \subseteq \mathcal{H}. \tag{B.20}$$

Observe that since $\mathcal{M} = \overline{M} \supseteq M$, then $\mathcal{M}^\perp \subseteq M^\perp$ and also $M^{\perp\perp} \subseteq \mathcal{M}^{\perp\perp} = \mathcal{M}$, from where it follows that $M^{\perp\perp} = \mathcal{M}$ since $\mathcal{M}$ is the *smallest* subspace containing M.

definition 14. Let $\mathcal{M}, \mathcal{N}$ be two subspaces of $\mathcal{H}$. The *algebraic sum* of $\mathcal{M}$ and $\mathcal{N}$, denoted by $\mathcal{M} + \mathcal{N}$ is the set of all vectors of the form $|\phi\rangle + |\psi\rangle$ with $|\phi\rangle \in \mathcal{M}, |\psi\rangle \in \mathcal{N}$.

The set $\mathcal{M} + \mathcal{N}$ is clearly a linear manifold, and indeed it is the smallest linear manifold containing both $\mathcal{M}$ and $\mathcal{N}$, i.e., (using the notation of definition 10),

$$\mathcal{M} + \mathcal{N} = [\mathcal{M} \cup \mathcal{N}]. \tag{B.21}$$

However, $\mathcal{M} + \mathcal{N}$ is *not*, in general, a subspace of $\mathcal{H}$ because it may happen that it is not closed.

definition 15. Let $\mathcal{M}$ and $\mathcal{N}$ be subspaces of $\mathcal{H}$. The closure of $\mathcal{M} + \mathcal{N}$, which is a subspace of $\mathcal{H}$ denoted by $\mathcal{M} \uplus \mathcal{N}$ is said to be the *algebraic union* of $\mathcal{M}$ and $\mathcal{N}$. We have

$$\mathcal{M} \uplus \mathcal{N} = \overline{[\mathcal{M} + \mathcal{N}]} \tag{B.22}$$

Definition 15 generalizes for any countable (even infinite) union of subspaces, i.e., if $\{\mathcal{M}_i\}$ is a countable (finite or infinite) family of subspaces of $\mathcal{H}$, we define their *algebraic union* by

$$\uplus_i \mathcal{M}_i = \overline{[\sum_i \mathcal{M}_i]} \tag{B.23}$$

where $\sum_i \mathcal{M}_i$ is the (algebraic) sum of the $\mathcal{M}_i$.

proposition 16. Let $\mathcal{M}, \mathcal{N}$ be subspaces of $\mathcal{H}$. The *set theoretical intersection* of $\mathcal{M}$ and $\mathcal{N}$, denoted as usual by $\mathcal{M} \cap \mathcal{N}$ is a subspace of $\mathcal{H}$. It is the *largest* subspace contained in both $\mathcal{M}$ and $\mathcal{N}$.

proposition 17. Let $\mathcal{H}_1, \mathcal{H}_2, \mathcal{H}_3 \subseteq \mathcal{H}$ be arbitrary subspaces. Then,

$$\begin{aligned}(\mathcal{H}_1 \cap \mathcal{H}_2) \cap \mathcal{H}_3 &= \mathcal{H}_1 \cap (\mathcal{H}_2 \cap \mathcal{H}_3) \\ (\mathcal{H}_1 \uplus \mathcal{H}_2) \uplus \mathcal{H}_3 &= \mathcal{H}_1 \uplus (\mathcal{H}_2 \uplus \mathcal{H}_3)\end{aligned} \tag{B.24}$$

definition 18. Let $\mathcal{M}, \mathcal{N}$ be subspaces of $\mathcal{H}$. If $\mathcal{M} \cap \mathcal{N} = \{|0\rangle\}$, the linear manifold $\mathcal{M} + \mathcal{N}$ is said to be the direct sum of $\mathcal{M}$ and $\mathcal{N}$ and is denoted by $\mathcal{M} \oplus \mathcal{N}$.

It is important to have in mind that $\mathcal{M} \oplus \mathcal{N}$ is not in general a subspace of $\mathcal{H}$. When $\mathcal{M}$ is a subspace of $\mathcal{H}$, then using the definition of $\mathcal{M}^\perp$ we get immediately that $\mathcal{M} \cap \mathcal{M}^\perp = \{|0\rangle\}$. Propositions 20, 21 and 22 below are important steps for proving that $\mathcal{H} = \mathcal{M} \oplus \mathcal{M}^\perp$.

definition 19. Let $|\phi\rangle \in \mathcal{H}$ and S be any subset of $\mathcal{H}$. The *distance* from $|\phi\rangle$ to S is defined by

$$d(|\phi\rangle, S) = \inf\{\| \, |\phi\rangle - |\psi\rangle \, \|\} : |\psi\rangle \in S, \qquad (B.25)$$

i.e., d is the greatest lower bound of the distances from $|\phi\rangle$ to the points of S.

proposition 20. If $\mathcal{M}$ is a proper subspace of $\mathcal{H}$ (i.e., $\mathcal{M} \neq |0\rangle, \mathcal{M} \neq \mathcal{H}$), then $\exists |\phi_0\rangle \in \mathcal{H}, |\phi_0\rangle \neq 0$ such that $|\phi_0\rangle \perp \mathcal{M}$ (i.e., $|\phi_0\rangle$ is orthogonal to all vectors of $\mathcal{M}$).

proposition 21. Let $\mathcal{M}, \mathcal{N}$ be subspaces of $\mathcal{H}$ such that $\mathcal{M} \perp \mathcal{N}$. Then, the linear manifold $\mathcal{M} + \mathcal{N}$ is a subspace.

proposition 22. If $\mathcal{M}$ is a subspace of $\mathcal{H}$, then $\mathcal{H} = \mathcal{M} \oplus \mathcal{M}^\perp$.

corollary 23. If $\mathcal{H} = \mathcal{M} \oplus \mathcal{M}^\perp$ then

$$\mathcal{M} \uplus \mathcal{M}^\perp = \mathcal{M} \oplus \mathcal{M}^\perp. \qquad (B.26)$$

Proposition 22 can easily be generalized for a sequence of mutually orthogonal subspaces $\{\mathcal{H}_i\}, i = 1, 2...$, of $\mathcal{H}$ ($\mathcal{H}_i \cap \mathcal{H}_j = \{|0\rangle\}, i \neq j, i, j = 1, 2...\}$, i.e., the sequence spans $\mathcal{H}$

$$\mathcal{H} = \oplus_{i=1}^{\infty} \mathcal{H}_i \qquad (B.27)$$

Equation (B.27) implies that $\forall |\phi\rangle \in \mathcal{H}$ can be written in a unique way as

$$|\phi\rangle = \sum_{i=1}^{\infty} |\phi_i\rangle, \ |\phi_i\rangle \in \mathcal{H}_i \qquad (B.28)$$

where the infinite sum means $\sum_{i=1}^{\infty} |\phi_i\rangle \to |\phi\rangle$.

definition 24. Let $\mathcal{H}_1$ be any subspace of $\mathcal{H}$. The (set) mapping

$$\perp : \mathcal{H}_1 \to \mathcal{H}_1^\perp \qquad (B.29)$$

is called *orthocomplementation*.

definition 25. Let $\mathcal{H}_1, \mathcal{H}_2$ be two subspaces of $\mathcal{H}$. They are said to be *compatible* and we denote this fact by writing $\mathcal{H}_1 \leftrightarrow \mathcal{H}_2$ if and only if

$$(\mathcal{H}_1 \cap \mathcal{H}_2) \uplus (\mathcal{H}_1 \cap \mathcal{H}_2^\perp) = \mathcal{H}_1. \qquad (B.30)$$

It is easy to show that from eq.(B.30) it follows that

$$(\mathcal{H}_2 \cap \mathcal{H}_1) \uplus (\mathcal{H}_2 \cap \mathcal{H}_1^\perp) = \mathcal{H}_2, \qquad (B.31)$$

which shows that compatibility is a symmetric relation, justifying the use of the symbol $\leftrightarrow$. However, in general, *the compatibility relation is not transitive*, i.e., if $\mathcal{H}_1, \mathcal{H}_2, \mathcal{H}_3$ are three subspaces of H and if $\mathcal{H}_1 \leftrightarrow \mathcal{H}_2, \mathcal{H}_2 \leftrightarrow \mathcal{H}_3$ it does not follows that $\mathcal{H}_1 \leftrightarrow \mathcal{H}_3$.

definition 26. Let $\mathcal{H}_1, \mathcal{H}_2$ be two subspaces of $\mathcal{H}$. We say that they are *disjoint* if $\mathcal{H}_1 \subseteq \mathcal{H}_2^\perp$ and we write $\mathcal{H}_1 \perp \mathcal{H}_2$.

Definition 26 implies that $\mathcal{H}_2 \subseteq \mathcal{H}_1^\perp$, showing that the disjoint relation expressed through the set inclusion operation $\subseteq$ is symmetric. This gives us another possibility of expressing the compatibility relation.

proposition 27. Let $\mathcal{H}_1, \mathcal{H}_2$ be two subspaces of $\mathcal{H}$. They are *compatible* if there exists three mutually disjoint subspaces $\mathcal{M}, \mathcal{N}$ and $\mathcal{K}$ such that

$$\mathcal{H}_1 = \mathcal{M} \uplus \mathcal{K}, \quad \mathcal{H}_2 = \mathcal{N} \uplus \mathcal{K}. \tag{B.32}$$

Observe that from eq.(B.32) it follows immediately that

$$\mathcal{K} = \mathcal{H}_1 \cap \mathcal{H}_2, \quad \mathcal{M} = \mathcal{H}_1 \cap \mathcal{K}^\perp, \quad \mathcal{N} = \mathcal{H}_2 \cap \mathcal{K}^\perp. \tag{B.33}$$

proposition 28. Let $\mathcal{H}_1, \mathcal{H}_2, \mathcal{H}_3$ be three subspaces of $\mathcal{H}$ which are pairwise compatible. Then they satisfy the *distributive* laws

$$\begin{aligned}\mathcal{H}_1 \cap (\mathcal{H}_2 \uplus \mathcal{H}_3) &= (\mathcal{H}_1 \cap \mathcal{H}_2) \uplus (\mathcal{H}_1 \cap \mathcal{H}_3), \\ \mathcal{H}_1 \uplus (\mathcal{H}_2 \cap \mathcal{H}_3) &= (\mathcal{H}_1 \uplus \mathcal{H}_2) \cap (\mathcal{H}_1 \uplus \mathcal{H}_3).\end{aligned} \tag{B.34}$$

The proof of relations (B.34) is trivial. It is very important to have in mind that given three subspaces $\mathcal{H}_1, \mathcal{H}_2, \mathcal{H}_3$ of $\mathcal{H}$, in general the distributive laws do not hold.

example 29. As an illustrative example, consider a two-dimensional Hilbert space $\mathcal{H}$ and two one-dimensional subspaces, $\mathcal{H}_1$ and $\mathcal{H}_2 = \mathcal{H}_1^\perp$ of $\mathcal{H}$. Let, moreover $\mathcal{H}_3 \neq \mathcal{H}_1, \mathcal{H}_2$ be an one dimensional subspace of $\mathcal{H}$. Then

$$\mathcal{H}_3 \cap (\mathcal{H}_1 \uplus \mathcal{H}_2) = \mathcal{H}_3 \cap \mathcal{H} = \mathcal{H}_3. \tag{B.35}$$

On the other hand, since

$$\mathcal{H}_3 \cap \mathcal{H}_1 = \{|0\rangle\} = \mathcal{H}_3 \cap \mathcal{H}_1^\perp, \tag{B.36}$$

it follows that

$$\mathcal{H}_3 \cap (\mathcal{H}_1 \uplus \mathcal{H}_2) \neq (\mathcal{H}_3 \cap \mathcal{H}_1) \uplus (\mathcal{H}_3 \cap \mathcal{H}_2). \tag{B.37}$$

proposition 30. Let $\mathcal{H}$ be a finite dimensional Hilbert space and let $\mathcal{H}_1, \mathcal{H}_2, \mathcal{H}_3$ be three subspaces of $\mathcal{H}$, such that $\mathcal{H}_1 \subseteq \mathcal{H}_3$. Then the so called *modular law* holds, i.e.,

$$\mathcal{H}_1 \uplus (\mathcal{H}_2 \cap \mathcal{H}_3) = (\mathcal{H}_1 \uplus \mathcal{H}_2) \cap \mathcal{H}_3. \tag{B.38}$$

The proof of eq.(B.38) can be done by verifying that $\mathcal{H}_1 \uplus (\mathcal{H}_2 \cap \mathcal{H}_3) \subseteq (\mathcal{H}_1 \uplus \mathcal{H}_2) \cap \mathcal{H}_3$ and that $(\mathcal{H}_1 \uplus \mathcal{H}_2) \cap \mathcal{H}_3 \subseteq \mathcal{H}_1 \uplus (\mathcal{H}_2 \cap \mathcal{H}_3)$.

proposition 31. Let $\mathcal{H}$ be an infinite dimensional Hilbert space and let $\mathcal{H}_1, \mathcal{H}_2, \mathcal{H}_3$ be three subspaces of $\mathcal{H}$, such that $\mathcal{H}_1 \subseteq \mathcal{H}_3$. Then we have in general that

$$\mathcal{H}_1 \uplus (\mathcal{H}_2 \cap \mathcal{H}_3) \subseteq (\mathcal{H}_1 \uplus \mathcal{H}_2) \cap \mathcal{H}_3. \tag{B.39}$$

Proof: We give a proof of proposition 31 for the simple case where $\mathcal{H}_1 \cap \mathcal{H}_2 = \{|0\rangle\}$. Let

$$|\phi\rangle \in \mathcal{H}_1 \uplus \mathcal{H}_2 = \overline{[\mathcal{H}_1 + \mathcal{H}_2]}, \tag{B.40}$$

and suppose that it is a limit of vectors of the form $|\phi_1\rangle + |\phi_2\rangle$, with $|\phi_1\rangle \in \mathcal{H}_1$, $|\phi_2\rangle \in \mathcal{H}_2$, but suppose that $|\phi\rangle$ itself *does not have* such a form. Let $\mathcal{H}_3$ be the subspace generated by $\mathcal{H}_1$ and $|\phi\rangle$. Then, $\forall |\psi\rangle \in \mathcal{H}_2 \cap \mathcal{H}_3$ is of the form

$$|\psi\rangle = c|\phi\rangle + |\phi_1\rangle, \ c \in \mathcal{C}, |\phi_1\rangle \in \mathcal{H}_1 . \tag{B.41}$$

Now, $|\psi\rangle \in \mathcal{H}_2$ and $|\phi_1\rangle \in \mathcal{H}_1$. But this is impossible unless $|\psi\rangle = |0\rangle$ and then $\mathcal{H}_2 \cap \mathcal{H}_3 = \{|0\rangle\}$. Then,

$$\mathcal{H}_1 \uplus (\mathcal{H}_2 \cap \mathcal{H}_3) = \mathcal{H}_1, \tag{B.42}$$

and

$$(\mathcal{H}_1 \uplus \mathcal{H}_2) \cap \mathcal{H}_3 = \mathcal{H}_3. \tag{B.43}$$

Eqs.(B.42-43) and the hypothesis $\mathcal{H}_1 \subseteq \mathcal{H}_3$ imply eq.(B.39).

B.2.1 THE LATTICE OF SUBSPACES

Recalling the definitions of lattice theory given in Appendix A and the properties proved above, we have the following:

proposition 32. The set of subspaces of an infinite dimensional Hilbert space $\mathcal{H}$ is an atomic, complete, orthocomplemented and non modular lattice. This lattice, called the *lattice of subspaces* of $\mathcal{H}$ will be denoted $\mathfrak{L}(\mathcal{H})$.

Proof: Let $\mathcal{L}$ be an abstract lattice with the properties stated above. All we need for proving the proposition is:

(i) to recall definition 44 of Appendix A of lattice isomorphism, $i : \mathcal{L} \to \mathfrak{L}(\mathcal{H})$,

(ii) to interpret the operations $\vee$, $\wedge$, $'(=\perp$, see definition 47 of Appendix A) and $\leq$ as the corresponding the operations $\uplus, \cap, \perp$ (see definition 24) and $\subseteq$ in $\mathfrak{L}(\mathcal{H})$.

corollary 33. If $\mathcal{H}$ is a finite dimensional Hilbert space, then $\mathfrak{L}(\mathcal{H})$ is a modular lattice.

B.3 DUAL AND CONJUGATE DUAL SPACES

definition 34. Let V be a linear (vector) space over a commutative field $\mathcal{K}$. The dual space of V, denoted by V^* is by definition the linear space (over $\mathcal{K}$) of all linear forms over V, i.e.,

$$V^* = \{F | F(\mathbf{v}) \in \mathcal{K}, \ \forall \mathbf{v} \in V\}. \tag{B.44}$$

The elements of V^* are usually called *covectors* or *forms* and if $F, F_1, F_2 \in V^*$, $\mathbf{v}, \mathbf{w} \in V$ and $k \in \mathcal{K}$, the following properties hold

$$\begin{align}
F(\mathbf{v}+\mathbf{w}) &= F(\mathbf{v}) + F(\mathbf{w}), \ F(k\mathbf{v}) = kF(\mathbf{v}), \tag{B.45} \\
(F_1 + F_2)(\mathbf{v}) &= F_1(\mathbf{v}) + F_2(\mathbf{v}), \ (kF)(\mathbf{v}) = k(F(\mathbf{v})). \tag{B.46}
\end{align}$$

For a Hilbert space $\mathcal{H}$ the covectors are also called *linear functionals* or *bra vectors* (or simply *bras*). We denote bras by symbols like $\langle F|$, $\langle x|$, etc... and write the action of $\langle F| \in \mathcal{H}^*$ on $|\phi\rangle \in \mathcal{H}$ as

$$\langle F| : H \to C, |\phi\rangle \mapsto \langle F|\phi\rangle. \tag{B.47}$$

definition 35. Let $\langle F| \in \mathcal{H}^*$. $\langle F|$ is said to be a *bounded bra* (or *bounded functional*) if $\forall |\phi\rangle \in \mathcal{H}$ there exists a real number $m < \infty$ such that

$$|\langle F|\phi\rangle| < m \, \| \, |\phi\rangle \, \|, \; m < \infty \tag{B.48}$$

definition 36. The *norm* of $\langle F| \in \mathcal{H}^*$ denoted $\| \, \langle F| \, \|$ is the greatest lower bound of all numbers m satisfying eq.(B.48).

proposition 37. Let $\langle \psi | \in \mathcal{H}^*$ be a bounded bra. Then, $\forall |\phi\rangle \in \mathcal{H}$ there exists a unique $|\psi\rangle \in \mathcal{H}$ such that

$$\langle \psi | \phi \rangle = (|\psi\rangle, |\phi\rangle), \tag{B.49}$$
$$\| \, \langle \psi | \, \| = \| \, |\psi\rangle \, \|. \tag{B.50}$$

Proposition 37 is known as the *Riesz representation theorem* and eq.(B.49) is the reason for the physicist's representation of the scalar product used in the main text.

observation 38. The set of bounded bras equipped with the scalar product

$$(\, , \,) : \mathcal{H}^* \times \mathcal{H}^* \to \mathcal{C},$$
$$(\langle \psi |, \langle \phi |) = (|\psi\rangle, |\phi\rangle) = \langle \psi | \phi \rangle. \tag{B.51}$$

is a *pre-Hilbert space*.

proposition 39. The closure in the norm (given by eq.(B.50)) of the pre-Hilbert space of all bounded bras, with the scalar product given by eq.(B.51) is a Hilbert space, which is called the *conjugate* (or *normed*) dual of $\mathcal{H}$ and denoted $\mathcal{H}^\dagger$.

observation 40. Let S be a differential manifold[3], e.g., S may be the Newtonian spacetime[2], or Minkowski spacetime[3,4], or the configuration space of a many particle system (see, e.g.,[7] of Appendix A.). Let $(S, \mathfrak{M}, \mu)$ be a measurable space and consider the realization of the abstract Hilbert space $\mathcal{H}$ (of infinite dimension) as $L^2(S, \mathfrak{M}, \mu)$ (see definition 89 of Appendix A). The elements of $L^2(S, \mathfrak{M}, \mu)$ are mappings usually denoted by ψ, but here also denoted conveniently by $|\psi\rangle$ such that

$$L^2(S, \mathfrak{M}, \mu) \ni |\psi\rangle : S \to \mathcal{C}; x \mapsto \psi(x), \tag{B.52}$$

and the scalar product and the norm are given $\forall |\psi\rangle, |\phi\rangle \in L^2(S, \mathfrak{M}, \mu)$, by

$$(\psi, \phi) = \langle \psi | \phi \rangle = \int \psi^*(x) \phi(x) d\mu,$$
$$\| \, \psi \, \| = \| \, |\psi\rangle \, \| = \sqrt{(\psi, \psi)} = \sqrt{\langle \psi | \psi \rangle}. \tag{B.53}$$

definition 41. The *Dirac measure* is the functional $\langle x | \in (L^2(S, \mathfrak{M}, \mu))^*$ such that $\forall |\psi\rangle \in L^2(S, \mathfrak{M}, \mu)$

$$\langle x | \psi \rangle = \psi(x). \tag{B.54}$$

It is clear that $\langle x | \in (L^2(S, \mathfrak{M}, \mu))^*$ *is not* a bounded bra, and since $\forall |\psi\rangle \in L^2(S, \mathfrak{M}, \mu)$ there is no $|x\rangle \in L^2(S, \mathfrak{M}, \mu)$ such that $\langle x | \psi \rangle = (|x\rangle, |\psi\rangle)$.[4]

B.4   TENSOR PRODUCTS

definition 42. Let V_i, $i = 1, 2, ..., n$ and V be vector spaces over a commutative field $\mathcal{K}$. A mapping

$$\mathbf{T} : V_1 \times V_2 \times ... \times V_n \to V, \tag{B.55}$$

is called *multilinear* if

(i) $\mathbf{T}(\mathbf{v}_1, ..., \mathbf{v}_i + \mathbf{v}'_i, ..., \mathbf{v}_n) = \mathbf{T}(\mathbf{v}_1, ..., \mathbf{v}_i, , ..., \mathbf{v}_n) + \mathbf{T}(\mathbf{v}_1, ..., \mathbf{v}'_i, ..., \mathbf{v}_n)$,

(ii) $\mathbf{T}(\mathbf{v}_1, ..., k\mathbf{v}_i, ..., \mathbf{v}_n) = k\mathbf{T}(\mathbf{v}_1, ..., \mathbf{v}_i, ..., \mathbf{v}_n)$, $k \in \mathcal{K}$. (B.56)

proposition 43. Let $L(V_1, V_2, ..., V_n; V)$ be the set of all multilinear forms from $V_1 \times V_2 \times ... \times V_n \to V$. $L(V_1, V_2, ..., V_n; V)$ has a natural linear (vector) space structure over $\mathcal{K}$ if we define for $\forall \mathbf{T}, \mathbf{Q} \in L(V_1, V_2, ..., V_n; V)$ and $k \in \mathcal{K}$,

$$\begin{aligned}(\mathbf{T} + \mathbf{Q})(\mathbf{v}_1, ..., \mathbf{v}_i, ..., \mathbf{v}_n) &= \mathbf{T}(\mathbf{v}_1, ..., \mathbf{v}_i, ..., \mathbf{v}_n) + \mathbf{Q}(\mathbf{v}_1, ..., \mathbf{v}_i, ..., \mathbf{v}_n), \\ (k\mathbf{T})(\mathbf{v}_1, ..., \mathbf{v}_i, ..., \mathbf{v}_n) &= k[\mathbf{T}(\mathbf{v}_1, ..., \mathbf{v}_i, ..., \mathbf{v}_n)].\end{aligned} \tag{B.57}$$

When all the $V_i = W$ we denote $L(W, W, ..., W; V)$ by $L^n(W; V)$ and $\mathbf{T} \in L^n(W; \mathcal{K})$ is called a *multilinear form* on W. In particular, $L^1(W; \mathcal{K}) = L(W; \mathcal{K}) = W^\star$ is the dual space of W.

definition 44. A *tensor product* of the linear spaces $V_1, V_2, ..., V_n$ is a linear space V together with a multilinear mapping denoted $\otimes$,

$$\begin{aligned}\otimes \; &: \; V_1 \times V_2 \times ... \times V_n \to V = V_1 \otimes V_2 \otimes ... \otimes V_n, \\ (\mathbf{v}_1, \mathbf{v}_2, ..., \mathbf{v}_n) &\longmapsto \mathbf{v}_1 \otimes \mathbf{v}_2 \otimes ... \otimes \mathbf{v}_n,\end{aligned} \tag{B.58}$$

which for any multilinear mapping $\mathbf{T} : V_1 \times V_2 \times ... \times V_n \to \mathcal{K}$ is such that there exists a unique mapping

$$\widehat{\mathbf{T}} : V_1 \otimes V_2 \otimes ... \otimes V_n \to \mathcal{K}.$$

for which $T = \widehat{\mathbf{T}} \circ \otimes$.

proposition 45. A tensor product of the linear spaces $V_1, V_2, ..., V_n$ always exists and any two tensor products are naturally isomorphic.

In view of proposition 45, we talk from now on of *the* tensor product.

proposition 46. The elements of $V_1 \otimes V_2 \otimes ... \otimes V_n$ are sums of scalar multiples of tensor products of vectors (like in eq.(B.58)), i.e.,

$$V_1 \otimes ... \otimes V_n \ni T = \underbrace{a(\mathbf{v}_1 \otimes ... \otimes \mathbf{v}_n) + a'(\mathbf{v}'_1 \otimes ... \otimes \mathbf{v}'_n) + (...)}_{\text{finite number of terms}}, \tag{B.59}$$

where $\mathbf{v}_i..., \mathbf{v}'_i \in V i, i = 1, 2, ..., n$ and $a, a' \in \mathcal{K}$, and the following properties hold

$$\begin{aligned}\mathbf{v}_1 \otimes ... \otimes (\mathbf{v}_i + \mathbf{v}'_i) \otimes ... \otimes \mathbf{v}_n &= \mathbf{v}_1 \otimes ... \otimes \mathbf{v}_i \otimes ... \otimes \mathbf{v}_n \\ &\quad + \mathbf{v}_1 \otimes ... \otimes \mathbf{v}'_i \otimes ... \otimes \mathbf{v}_n, \\ (a\mathbf{v}_1) \otimes ... \otimes \mathbf{v}'_i \otimes ... \otimes \mathbf{v}_n &= ... = \mathbf{v}_1 \otimes ... \otimes (a\mathbf{v}'_i) \otimes ... \otimes \mathbf{v}_n = ... \\ &= a(\mathbf{v}_1 \otimes ... \otimes \mathbf{v}'_i \otimes ... \otimes \mathbf{v}_n).\end{aligned} \tag{B.60}$$

proposition 47. The *tensor product of Hilbert spaces* $\mathcal{H}_i, i = 1, 2, ..., n$, is a Hilbert space denoted $\mathcal{H}_1 \otimes \mathcal{H}_2 \otimes ... \otimes \mathcal{H}_n =^\otimes \mathcal{H}$ if the multilinear mapping

$$\otimes : \mathcal{H}_1 \times \mathcal{H}_2 \times ... \times \mathcal{H}_n \to \mathcal{H}_1 \otimes \mathcal{H}_2 \otimes ... \otimes \mathcal{H}_n,$$
$$(|\psi_1\rangle, |\psi_2\rangle, ..., |\psi_n\rangle) \mapsto |\psi_1\rangle \otimes |\psi_2\rangle \otimes ... \otimes |\psi_n\rangle, \tag{B.61}$$

is such that:

(i) the set of all vectors $\{|\varphi_{1,i_1}\rangle \otimes |\varphi_{2,i_2}\rangle \otimes ... \otimes |\varphi_{n,i_n}\rangle\} = \{\otimes_{j,i_j} |\varphi_{j,i_j}\rangle\}$ where $\{|\varphi_{j,i_j}\rangle\}$, $j = 1, 2, ..., n$, $i_j = 1, 2, ...$ are complete orthonormal sets for $\mathcal{H}_j$, $j = 1, 2, ..., n$ spans $^\otimes \mathcal{H}$, i.e., $\forall |\Psi\rangle \in^\otimes \mathcal{H}$ can be written as

$$|\Psi\rangle = \sum_{i_1...i_n} \psi_{i_1...i_n} |\varphi_{1,i_1}\rangle \otimes |\varphi_{2,i_2}\rangle \otimes ... \otimes |\varphi_{n,i_n}\rangle; \tag{B.62}$$

(ii) the *scalar product* in $^\otimes \mathcal{H}$ is the mapping

$$(\ ,\ ) :^\otimes \mathcal{H} \times^\otimes \mathcal{H} \to \mathcal{C}$$
$$(|\Psi\rangle, |\Phi\rangle) = \langle \Psi | \Phi \rangle$$
$$= \sum_{i_1...i_n, j_1...j_n} \psi^*_{i_1...i_n} \phi_{j_1...j_n} \langle \varphi_{1,i_1} | \varphi_{1,j_1} \rangle ... \langle \varphi_{n,i_n} | \varphi_{n,j_n} \rangle$$
$$= \sum_{i_1...i_n} \psi^*_{i_1...i_n} \phi_{i_1...i_n}; \tag{B.63}$$

(iii) the norm of $|\Psi\rangle \in^\otimes \mathcal{H}$ is given by

$$\| |\Psi\rangle \| = \sqrt{\langle \Psi | \Psi \rangle} \geq 0. \tag{B.64}$$

The equality sign in eq.(B.64) holds only if $|\Psi\rangle = |0\rangle$.

The proof of proposition 47 is the simple verification that $^\otimes \mathcal{H}$ is indeed separable and complete. Moreover, the notation for the scalar product in the second line of eq.(B.63) implies that $\langle \Psi| \in^\otimes \mathcal{H}^\dagger$, where $^\otimes \mathcal{H}^\dagger$ is the conjugate dual of $^\otimes \mathcal{H}$.

B.5 LINEAR OPERATORS

definition 48. Let $\mathcal{H}_1, \mathcal{H}_2$ be Hilbert spaces. The linear space $\mathcal{H}_1 \otimes \mathcal{H}_2^\dagger$ is the set of all linear mappings (*operators*) from $\mathcal{H}_2$ to $\mathcal{H}_1$, i.e.,

$$\mathcal{H}_1 \otimes \mathcal{H}_2^\dagger \ni \mathbf{T} : \mathcal{H}_2 \to \mathcal{H}_1. \tag{B.65}$$

definition 49. A *bounded linear* operator on $\mathcal{H}$ is a mapping

$$\mathcal{H} \otimes \mathcal{H}^\dagger \ni \mathbf{O} : \mathcal{H} \supseteq D_\mathbf{O} \to R_\mathbf{O} \subseteq \mathcal{H} \tag{B.66}$$

where $D_\mathbf{O}$ (the *domain* of $\mathbf{O}$) and $R_\mathbf{O}$ (the *range* or *codomain* of $\mathbf{O}$) are linear manifolds, not necessarily subspaces of $\mathcal{H}$, and such that

$$\mathbf{O}(|\phi\rangle + |\psi\rangle) = \mathbf{O}|\phi\rangle + \mathbf{O}|\psi\rangle, \forall |\phi\rangle, |\psi\rangle \in D_\mathbf{O},$$
$$\mathbf{O}(c|\phi\rangle) = c\mathbf{O}|\phi\rangle), \forall |\phi\rangle \in D_\mathbf{O}, \forall c \in \mathcal{C},$$
$$\| \mathbf{O}|\phi\rangle \| = m \| |\phi\rangle \|, 0 \leq m < \infty, \forall |\phi\rangle \in D_\mathbf{O}. \tag{B.67}$$

The greatest lower bound of the numbers m is called the norm of **O** and is denoted by $\|\mathbf{O}\|$. If an m as above does not exist, then **O** is said to be *unbounded*. Unbounded operators which exist only when $\mathcal{H}$ is *infinite dimensional* are fundamental in functional analysis and quantum theory, as, e.g., the so-called *creation* and *annihilation* operators of quantum field theory (see[7] of Appendix A). For an unbounded operator $\mathbf{K} \in \mathcal{H} \otimes \mathcal{H}^\dagger$ there always exists a $|\phi\rangle \in \mathcal{H}$ such that for any finite real number m we have

$$\|\mathbf{K}|\phi\rangle\| > m \||\phi\rangle\|. \tag{B.68}$$

definition 50. An operator **O** is said to be *continuous* at $|\psi\rangle \in D_\mathbf{O}$ if the sequence $\{|\psi_i\rangle\}, i = 1, 2, ..., |\psi_i\rangle \in$ is such that $|\psi_i\rangle \to |\psi\rangle$ implies that $\mathbf{O}|\psi_i\rangle \to \mathbf{O}|\psi\rangle$.

proposition 51. Every bounded linear operator **O** on $\mathcal{H}$ is continuous everywhere and conversely, every linear operator continuous at $|\psi\rangle \in D_\mathbf{O}$ is continuous everywhere and bounded.

definition 52. Let $\mathbf{O}, \mathbf{O}_1 \in \mathcal{H} \otimes \mathcal{H}^\dagger$ be two operators with domains $D_\mathbf{O}$ and $D_{\mathbf{O}_1}$ with $D_\mathbf{O} \subseteq D_{\mathbf{O}_1}$ and such that $\mathbf{O}_1$ agrees with **O** on $D_\mathbf{O}$, i.e., $\mathbf{O}_{1|D_\mathbf{O}} = \mathbf{O}$. Then we say that $\mathbf{O}_1$ is the *extension* of O and we write $\mathbf{O} \subseteq \mathbf{O}_1$ or $\mathbf{O}_1 \supseteq \mathbf{O}$.

definition 53. Let $\{|\varphi_i\rangle\}, i = 1, 2..., |\varphi_i\rangle \in D_\mathbf{O}$ be a Cauchy sequence such that $|\varphi_i\rangle \to |\varphi\rangle$. Then, if $|\varphi\rangle \notin D_\mathbf{O}$ we define

$$\mathbf{O}|\varphi\rangle = \lim_{i \to \infty} \mathbf{O}|\varphi_i\rangle. \tag{B.69}$$

The limit in eq.(B.69) always exists since $\mathbf{O}|\varphi_i\rangle$ is a Cauchy sequence. Hence,

$$\|\mathbf{O}|\varphi_j\rangle - \mathbf{O}|\varphi_i\rangle\| \leq \|\mathbf{O}\| \||\varphi_j\rangle - |\varphi_i\rangle\|. \tag{B.70}$$

With definition 53 we obtain an *extension* of the linear bounded operator **O** from $D_\mathbf{O}$ to the closure $\overline{D}_\mathbf{O}$, which means that every bounded linear operator can be assumed as acting on subspaces of $\mathcal{H}$. If it happens that $\overline{D}_\mathbf{O} = \mathcal{H}$ (i.e., $D_\mathbf{O}$ is dense on $\mathcal{H}$), then the bounded linear operator acts on all $\mathcal{H}$.

It is *worthwhile* observing that even if $\overline{D}_\mathbf{O} \neq \mathcal{H}$, **O** can be extended to all $\mathcal{H}$. This can be done first by extending **O** to $\overline{D}_\mathbf{O}$ using definition 53 and then defining **O** in $D_\mathbf{O}^\perp$, as an arbitrarily bounded linear operator, e.g., putting $\mathbf{O}|\psi\rangle = |0\rangle, \forall |\psi\rangle \in D_\mathbf{O}^\perp$. Since $\mathcal{H} = D_\mathbf{O} \oplus D_\mathbf{O}^\perp$ the action of **O** on an arbitrary ket $|\phi\rangle \in \mathcal{H}$ is given by linearity.

Unless otherwise stated, all linear bounded operators used in what follows and in the main text are supposed to act on all $\mathcal{H}$.

definition 54. A *sesquilinear functional* is an element of $\mathcal{H}^\dagger \otimes \mathcal{H}^\dagger$ whose domain is $\mathcal{H} \otimes \mathcal{H}$.

definition 55. Any bounded linear operator $\mathbf{O} \in \mathcal{H} \otimes \mathcal{H}^\dagger$ defines a sesquilinear functional $\mathbf{G} \in \mathcal{H}^\dagger \otimes \mathcal{H}^\dagger$ by

$$\mathbf{G}(|\phi\rangle, |\psi\rangle) = (|\phi\rangle, \mathbf{O}|\psi\rangle) = \langle\phi|\mathbf{O}|\psi\rangle, \quad |\phi\rangle, |\psi\rangle \in \mathcal{H}. \tag{B.71}$$

definition 56. Each $\mathbf{G} \in \mathcal{H}^\dagger \otimes \mathcal{H}^\dagger$ defines *two* bounded linear operators on $\mathcal{H}$, **O** and $\mathbf{O}^\dagger$ which are said to be the *adjoint* to each other. We have,

$$\mathbf{G}(|\phi\rangle, |\psi\rangle) = (\mathbf{O}^\dagger|\phi\rangle, |\psi\rangle) = \langle\psi|\mathbf{O}^\dagger|\phi\rangle^* = \langle\phi|\mathbf{O}|\psi\rangle, \quad |\phi\rangle, |\psi\rangle \in \mathcal{H}. \tag{B.72}$$

definition 57. A bounded linear operator $\mathbf{O}$ such that $\mathbf{O} = \mathbf{O}^\dagger$ is said to be *self-adjoint* or *symmetrical* or *Hermitian*.

proposition 58. If $\mathbf{O}, \mathbf{O}_1, \mathbf{O}_2 \in \mathcal{H} \otimes \mathcal{H}^\dagger$, if $\mathbf{O}^\dagger, \mathbf{O}_1^\dagger, \mathbf{O}_2^\dagger \in \mathcal{H} \otimes \mathcal{H}^\dagger$ are respectively their adjoints and if $c \in \mathcal{C}$ we have,

$$\begin{aligned}(\mathbf{O}_1 + \mathbf{O}_2)^\dagger &= \mathbf{O}_1^\dagger + \mathbf{O}_2^\dagger, \\ (\mathbf{O}^\dagger)^\dagger &= \mathbf{O}, \\ (c\mathbf{O})^\dagger &= c^*\mathbf{O}_1, \\ (\mathbf{O}_1\mathbf{O}_2)^\dagger &= \mathbf{O}_2^\dagger\mathbf{O}_1^\dagger.\end{aligned} \quad (B.73)$$

Besides the self-adjoint operators, which are the representatives of observables in the formalism of quantum theory, there are other important kinds of linear bounded operators which are used in that theory, e.g.:

(i) the *projection operators* $\mathbf{P} \in \mathcal{H} \otimes \mathcal{H}^\dagger$ with domain $\mathcal{H}$ and range $R_\mathbf{P}$ which is always a subspace of $\mathcal{H}$ defined by

$$\mathbf{P}\mathbf{P}^\dagger = \mathbf{P}. \quad (B.74)$$

Also, given any subspace $\mathcal{H}_1 \subseteq \mathcal{H}$ there is a unique projection $\mathbf{P}_1$ such that $R_{\mathbf{P}_1} = \mathcal{H}_1$. If $\mathcal{H} = (\mathcal{H}_1 + \mathcal{H}_1^\perp) \ni |\phi\rangle = |\phi_1\rangle + |\phi_2\rangle$ then for a projection $\mathbf{P}_1$ such that $R_{\mathbf{P}_1} = \mathcal{H}_1$ we have

$$\mathbf{P}_1|\phi\rangle = |\phi_1\rangle. \quad (B.75)$$

(ii) the *unitary operators* $\mathbf{U} \in \mathcal{H} \otimes \mathcal{H}^\dagger$ which satisfy

$$\mathbf{U}\mathbf{U}^\dagger = \mathbf{U}^\dagger\mathbf{U} = \mathbf{1}, \quad (B.76)$$

where $\mathbf{1}$ is the identity operator in $\mathcal{H}$.

definition 59. The matrix representation of a linear bounded operator $\mathbf{O} \in \mathcal{H} \otimes \mathcal{H}^\dagger$ is given by the set of numbers (O_{ij}), such that

$$\begin{aligned}\mathbf{O} \mid \varphi_i\rangle &= \sum_j O_{ij}|\varphi_j\rangle, \\ O_{ij} &= \langle\varphi_j|\mathbf{O}|\varphi_i\rangle,\end{aligned} \quad (B.77)$$

where $\{|\varphi_i\rangle\}$ is an orthonormal basis of $\mathcal{H}$.

definition 60. A linear operator $\mathbf{O} \subseteq \mathbf{O}^{\dagger\dagger} \subseteq \mathbf{O}^\dagger$ is said to be *closed* if for every sequence $\{|\varphi_i\rangle\}$, $|\varphi_i\rangle \in D_\mathbf{O} \subseteq \mathcal{H}$ such that

$$|\varphi_i\rangle \to |\varphi\rangle, \; \mathbf{O}|\varphi_i\rangle \to |\psi\rangle,$$

we have

$$|\psi\rangle \in D_\mathbf{O}, \text{ and } \mathbf{O}|\varphi\rangle \to |\psi\rangle. \quad (B.78)$$

observation 61. To be closed and to be continuous are different properties that a given $\mathbf{O} \in \mathcal{H} \otimes \mathcal{H}^\dagger$ may or may not possess. Indeed, it is easy to find examples of closed operators that are not continuous and also it is easy to prove that every continuous operator is closed. Also it can be proved that every non closed operator can be extended to a closed one which is unique.

proposition 62. If $\mathbf{K} \in \mathcal{H} \otimes \mathcal{H}^\dagger$ is closed then $\mathbf{K}$ is necessarily bounded.

proposition 63. A closed and unbounded operator cannot be defined in all $\mathcal{H}$, i.e., if $\mathbf{O} \in \mathcal{H} \otimes \mathcal{H}^\dagger$ is closed and unbounded, then $D_\mathbf{O} \subset \mathcal{H}$.

definition 64. Let $\mathbf{O}$ be a closed and unbounded operator and suppose that $D_\mathbf{O}$ is dense in $\mathcal{H}$ and let $|\psi\rangle \in D_\mathbf{O}$. The adjoint $\mathbf{O}^\dagger$ of $\mathbf{O}$ is defined if there exists $|\phi\rangle \in D_{\mathbf{O}^\dagger}$ such that

$$(|\phi\rangle, \mathbf{O}|\psi\rangle) = (\mathbf{O}^\dagger|\phi\rangle, |\psi\rangle). \tag{B.79}$$

definition 65. Let $\mathbf{O}$ be an unbounded operator and suppose that $D_\mathbf{O}$ is dense in $\mathcal{H}$. $\mathbf{O}$ is said to be *symmetrical* (or *Hermitian*) if $\mathbf{O}^\dagger \supseteq \mathbf{O}$, $\mathbf{O}$ is said to be self-adjoint (and thus symmetrical) if $\mathbf{O}^\dagger = \mathbf{O}$.

proposition 66. For unbounded symmetrical operators, the following properties hold:

(i) $\mathbf{O} \subseteq \mathbf{O}^{\dagger\dagger} \subseteq \mathbf{O}^\dagger$,

(ii) $\mathbf{O} \subseteq \mathbf{O}^{\dagger\dagger}$ and $\mathbf{O}^{\dagger\dagger} \subset \mathbf{O}^\dagger \Rightarrow \mathbf{O}$ is closed and is not self-adjoint,

(iii) $\mathbf{O} \subset \mathbf{O}^{\dagger\dagger} = \mathbf{O}^\dagger \Rightarrow \mathbf{O}$ is not closed. $\mathbf{O}^{\dagger\dagger}$, its smallest closed extension is self-adjoint (it is called essentially self-adjoint),

(iv) $\mathbf{O} \subset \mathbf{O}^{\dagger\dagger} \subset \mathbf{O}^\dagger \Rightarrow \mathbf{O}$ is neither closed nor essentially self-adjoint,

(v) $\mathbf{O} = \mathbf{O}^{\dagger\dagger} = \mathbf{O}^\dagger \Rightarrow \mathbf{O}$ is self-adjoint and closed.

We end this section with the remark that since for closed unbounded operators $\mathbf{O}_1, \mathbf{O}_2 \in \mathcal{H} \otimes \mathcal{H}^\dagger$ in general $\mathcal{H} \supseteq D_{\mathbf{O}_1} \neq R_{\mathbf{O}_2} \subseteq \mathcal{H}$ (and also in general $D_{\mathbf{O}_1} \neq D_{\mathbf{O}_2} \neq R_{\mathbf{O}_1}$) the composition (given by the product) of $\mathbf{O}_1$ and $\mathbf{O}_2$ and also the composition of $\mathbf{O}_2$ and $\mathbf{O}_1$, may be not defined. Finally, when the composition of linear operators is well defined the linear space of the set of operators close an algebra, the so called *operator algebra*, which is extensively used in quantum theory. There, we have to use Lie algebras[6] where the product of two of its elements which are linear operators on $\mathcal{H}$, let us say $\mathbf{A}$ and $\mathbf{B}$ is defined by the *commutator product*

$$[\mathbf{A}, \mathbf{B}] = \mathbf{AB} - \mathbf{BA} \tag{B.80}$$

B.6 PROJECTION OPERATORS AND LATTICE STRUCTURE

We have already defined projection operators as being bounded linear operators $\mathbf{P} \in \mathcal{H} \otimes \mathcal{H}^\dagger$, $D_\mathbf{P} = \mathcal{H}$ such that $\mathbf{PP}^\dagger = \mathbf{P}$, an equation that implies that $\mathbf{P}^2 = \mathbf{P}$ and $\mathbf{P} = \mathbf{P}^\dagger$, i.e., projection operators are idempotent and Hermitian.

Let $\mathcal{H}_1 \subseteq \mathcal{H}$ be a one dimensional subspace generated by $|\varphi\rangle \in \mathcal{H}$, i.e., $\mathcal{H}_1 = \{c|\varphi\rangle, c \in \mathcal{C}\}$. The projection on $\mathcal{H}_1$ is given by

$$\mathbf{P}_1 = \frac{|\varphi\rangle\langle\varphi|}{\langle\varphi|\varphi\rangle} = \frac{(c|\varphi\rangle)(\langle\varphi|c^*)}{\langle c\varphi|c\varphi\rangle} \tag{B.81}$$

This equation means that there is a *one-to-one* correspondence between $\mathbf{P}_1$ and $\mathcal{H}_1$. More generally, we can show without difficulties that given an indexed family $\{\mathbf{P}_\alpha\}$ the ranges $\mathbf{P}_\alpha$ are $R_{\mathbf{P}_\alpha} = \mathcal{H}_\alpha \subseteq \mathcal{H}$ where $\{\mathcal{H}_\alpha\}$ is an indexed family of subspaces. Conversely, to any arbitrary indexed family of subspaces $\{\mathcal{H}_\beta\}$, there corresponds an indexed family of projections $\{\mathcal{H}_\beta\}$ such that $R_{\mathbf{P}_\beta} = \mathcal{H}_\beta \subseteq \mathcal{H}$.

As a consequence of this one-to-one correspondence the lattice structure $\mathfrak{L}(\mathcal{H})$ of the subspaces of $\mathcal{H}$ (see section B.1 of this Appendix) can be transferred to the set of all projection operators.

Let us see how this can be done.

proposition 67. The set of all projection operators is a poset.

Proof: If $\{\mathcal{H}_\alpha\}$ is an indexed family of subspaces and $\{\mathbf{P}_\alpha\}$ the associated family of projection operators we can define the order relation by writing

$$P_\alpha \leq P_\beta \Leftrightarrow \mathcal{H}_\alpha \subseteq \mathcal{H}_\beta, \tag{B.82}$$

which proves the proposition.

Observe also that :

(i) since $\mathcal{H}_\alpha = R_{\mathbf{P}_\alpha} \subseteq R_{\mathbf{P}_\beta} = \mathcal{H}_\beta$, it follows that

$$\mathbf{P}_\alpha \mathbf{P}_\beta = \mathbf{P}_\alpha, \tag{B.83}$$

(ii) If $\mathbf{P}_\alpha$ is the projection on $\mathcal{H}_\alpha$, then $\mathbf{1} - \mathbf{P}_\alpha$ is clearly the projection on $\mathcal{H}_\alpha^\perp$.

In section B.1 of this Appendix, we identified in $\mathfrak{L}(\mathcal{H})$ the lattice operations $\wedge$ and $\vee$ respectively with the set theoretical intersection of subspaces (denoted $\cap$) and the algebraic union of subspaces (denoted $\uplus$). We now present how the lattice operations can be expressed in terms of algebraic operations involving projection operators.

proposition 68. Let $\{\mathcal{H}_\alpha\}$ be an indexed family of subspaces and $\{\mathbf{P}_\alpha\}$ the associated family of projection operators. Then:

(i) if $[\mathbf{P}_\alpha, \mathbf{P}_\beta] = 0$, and denoting $\mathbf{P}_{\alpha \wedge \beta} = \mathbf{P}_\alpha \wedge \mathbf{P}_\beta$ the projection corresponding to the intersection $\mathcal{H}_\alpha \cap \mathcal{H}_\beta$, we have

$$\mathbf{P}_{\alpha \wedge \beta} = \mathbf{P}_\alpha \mathbf{P}_\beta. \tag{B.84}$$

(ii) If $[\mathbf{P}_\alpha, \mathbf{P}_\beta] \neq 0$, then

$$\mathbf{P}_{\alpha \wedge \beta} = \mathbf{P}_\alpha \wedge \mathbf{P}_\beta = \lim_{n \to \infty} (\mathbf{P}_\alpha \mathbf{P}_\beta)^n \tag{B.85}$$

Proof: (i) is trivial. We now give a proof of (ii) for the particular case when $\mathcal{H}_\alpha$ and $\mathcal{H}_\beta$ are one dimensional *nonorthogonal* subspaces. To see that eq.(B.85) is correct for this case, consider the subspaces $\mathcal{H}_\alpha$ and $\mathcal{H}_\beta$ generated respectively by $|\psi_\alpha\rangle, |\psi_\beta\rangle \in \mathcal{H}$. Then,

$$\mathbf{P}_\alpha = \frac{|\psi_\alpha\rangle\langle\psi_\alpha|}{\langle\psi_\alpha|\psi_\alpha\rangle}, \quad \mathbf{P}_\beta = \frac{|\psi_\beta\rangle\langle\psi_\beta|}{\langle\psi_\beta|\psi_\beta\rangle}, \tag{B.86}$$

and we get immediately that

$$\lim_{n \to \infty} (\mathbf{P}_\alpha \mathbf{P}_\beta)^n |\Phi\rangle = |0\rangle, \forall |\Phi\rangle \in \mathcal{H}. \tag{B.87}$$

Also, if $[\mathbf{P}_\alpha, \mathbf{P}_\beta] = 0$ we have $(\mathbf{P}_\alpha \mathbf{P}_\beta)^n = \mathbf{P}_\alpha \mathbf{P}_\beta$ and we have eq.(B.84) again.

The lattice operation $\vee$ for the set of all projection operators will be denoted by the same symbol. Consider then the subspace $\mathcal{H}_\alpha \uplus \mathcal{H}_\beta$, the corresponding projection on this subspace denoted by $\mathbf{P}_{\alpha \vee \beta}$ is a *well defined* operation and we can verify that the following rule is the appropriate one

$$\mathbf{P}_{\alpha \vee \beta} = \mathbf{P}_\alpha \vee \mathbf{P}_\beta. \tag{B.88}$$

The algebraic expression for $\mathbf{P}_\alpha \vee \mathbf{P}_\beta$ can, in general, be calculated without difficulties. For the important case when $[\mathbf{P}_\alpha, \mathbf{P}_\beta] = 0$, it is easy to verify that

$$\mathbf{P}_{\alpha \vee \beta} = \mathbf{P}_\alpha \vee \mathbf{P}_\beta = \mathbf{P}_\alpha + \mathbf{P}_\beta - \mathbf{P}_\alpha \mathbf{P}_\beta. \tag{B.89}$$

For the algebraic union of an indexed family of subspaces eq.(B.87) is *generalized* in an obvious way.

With the above *identifications*, it follows that the set of all projection operators (which is a vector space, since it is a subset of $\mathcal{H} \otimes \mathcal{H}^\dagger$) has the *same* lattice structure (an isomorphism) as the set of subspaces of a given Hilbert space $\mathcal{H}$. In what follows, we denote the lattice generated by the set of all projection operators by $\mathfrak{L}^\mathbf{P}(\mathcal{H})$ and $\mathfrak{L}^\mathbf{P}(\mathcal{H}) \sim \mathfrak{L}(\mathcal{H})$.

proposition 69. Let $\mathfrak{L}^\mathbf{P}_\mathfrak{B}(\mathcal{H}) \subseteq \mathfrak{L}^\mathbf{P}(\mathcal{H})$ be an indexed family $\{\mathbf{P}_i\}$ of *commuting* projection operators which are:

(i) *closed* with respect to countable intersections ($\wedge$) and unions ($\vee$), i.e.,

$$\wedge_i \mathbf{P}_i \in \mathfrak{L}^\mathbf{P}_\mathfrak{B}(\mathcal{H}) \text{ and } \vee_i \mathbf{P}_i \in \mathfrak{L}^\mathbf{P}_\mathfrak{B}(\mathcal{H}),$$

(ii) $\forall \mathbf{P}_i \in \mathfrak{L}^\mathbf{P}_\mathfrak{B}(\mathcal{H}) \Rightarrow (\mathbf{1} - \mathbf{P}_i) \in \mathfrak{L}^\mathbf{P}_\mathfrak{B}(\mathcal{H})$,
(iii) $\mathbf{0} \in \mathfrak{L}^\mathbf{P}_\mathfrak{B}(\mathcal{H}), \mathbf{1} \in \mathfrak{L}^\mathbf{P}_\mathfrak{B}(\mathcal{H})$.

Then $\mathfrak{L}^\mathbf{P}_\mathfrak{B}(\mathcal{H})$ is a Boolean algebra.

These Boolean algebras are fundamental in quantum theory, since they correspond to the compatible propositions of the so called *proposition calculus* of quantum theory which is discussed in Chapter 7.

Before concluding this Appendix we shall show how to construct Boolean algebras for sets of commuting projections for the case where the Hilbert space $\mathcal{H}$ is identified with $L^2(S, \mathfrak{M}, \mu)$ as in observation 40 above.

Let $A \in \mathfrak{M}$, χ_A the characteristic function of A and consider any $\psi \in L^2(S, \mathfrak{M}, \mu)$. To the measurable set A we associate a projection operator $\mathbf{P}_A$ by

$$(\mathbf{P}_A \psi)(x) = \chi_A(x) \psi(x), \forall x \in S. \tag{B.90}$$

proposition 70. The projections $\mathbf{P}_A$ are σ-additive.

This follows immediately from our discussion in Appendix A, since $\mathfrak{M}$ is σ-additive. Then, if $\{A_i\}$ is a sequence of sets such that $A_i \cap A_j = \emptyset$, we have

$$\sum_{i=1}^{\infty} \mathbf{P}_{A_i} = \mathbf{P}_{\cup_{i=1}^\infty A_i},$$
$$\mathbf{P}_{A_i} \mathbf{P}_{A_j} = \mathbf{P}_{A_i \cup A_j}, \mathbf{P}_{A_i} \mathbf{P}_{A_j} = \mathbf{P}_{A_i \cap A_j},$$
$$\mathbf{P}_\emptyset = \mathbf{0}, \mathbf{P}_{A_i} = \mathbf{1}. \tag{B.91}$$

definition 71. A mapping

$$\mathbf{P} : \mathfrak{M} \to \mathfrak{L}_{\mathfrak{B}}^{\mathbf{P}}(\mathcal{H}), \ A \mapsto \mathbf{P}_A, \tag{B.92}$$

satisfying the properties in eq.(B.90) is said to be a *spectral measure* over $(S, \mathfrak{M}, \mu)$.

Given a spectral measure, it is possible to associate to it several different numerical measures. Particularly important for applications in quantum theory are the ones given by

$$\mu_\psi(A) = \langle \psi | \mathbf{P}_A | \psi \rangle. \tag{B.93}$$

Notes

1. In this book, when V is a Hilbert space (definition1), the vectors are denoted by *kets* like $|a\rangle, |f\rangle, |\phi\rangle$, etc. In this case the null vector is denoted by $|0\rangle$.

2. In the case of a non commutative field, like e.g., the quaternionic field used in quaternionic quantum theory, it is necessary to distinguish between a *left* and a *right* scalar multiplication. See[1] for details.

3. A good reference for a first study on differential manifolds is [6] of Appendix A.

4. To give more details on this topic it is necessary to introduce the concept of *rigged* Hilbert spaces, which is fundamental for certain problems of analysis and the study of quantum field theory. This concept will not be given here since it is not necessary for the main text. The interested reader is invited to consult [7] of Appendix A.

References

[1] S. De Leo and W. A. Rodrigues, Jr, Quantum mechanics: from complex to complexified quaternions, *Int. J. Theor. Phys.* **36**(12), 2725–2575 (1997).

[2] W. A. Rodrigues, Jr., Q. A. G. de Souza and Y. Bozhkov, The mathematical structure of Newtonian spacetime: classical dynamics and gravitation, *Found. Phys.* **25**(6), 871–924 (1995).

[3] W. A. Rodrigues, Jr. and M. A. F. Rosa, The meaning of time in the theory of relativity and Einstein's later view of the twin paradox, *Found. Phys.* **19**(6), 705–727 (1989).

[4] R. K. Sachs and H. Wu, *General Relativity for Mathematicians* (Springer-Verlag, Berlin, 1977).

[5] C. T. J. Dodson and T. C. Post, *Tensor Geometry* (Springer-Verlag, Berlin, 1991).

Index

absolute ego, 53
actual infinity, 62
actualization, 154
Adler, S., 7
Aharonov, Y., 41, 181, 184
Aharonov-Bohm effect, 118, 181, 182
Alexandrov, A. D., 141
algebra
    Clifford, 112
    σ-*Boolean*, 72
algebraic union, 204
and, *see* conjunction
antisymmetric wave function, 182
anyons, 182
approximation
    Born-Oppenheimer, 16
    WKB, 19
Aristotle, 67
Aspect, A., 4, 41, 101, 112, 113, 180
    experiment, 183
atomistic, 190
automaton
    nonoriented, 94
    normalized, 86, 88
averages, 120
Avogadro, 144
awareness, 51
axiom
    von Neumann's, 104

Badurek, G., 104, 114
Ballentine, L. E., 25, 28, 54, 55, 141
Barashenkov, V., 55, 56
barriers, 178
Barrow, J., 57
Barut, A. O., 121, 122, 131
Bastin, E., 55
Bauer, F., 50, 57, 63
Baylis, W. E., 112, 114, 130
BBT, *see* theory, de Broglie-Bohm *HVT*

beable(s), 37, 44, 91, 98, 115, 124, 134
Beauregard, O. C. de, 51, 57, 95, 128, 130
Belinfante, F. J., 98, 99, 101, 113
Bell, J. S., 35, 41, 42, 44, 55, 60, 64, 144, 148, 183
    inequalities, 1, 2, 35, 40, 46, 52, 83, 91, 94, 95, 160, 168, 183
    breakdown of, 91, 92, 95
    d'Espagnat form, 91, 135, 139
    original form of, 40
    original proof of, 39
    version of *MWI*, 59
Bennett, C. H., 168, 173, 179
Bernoulli theorem, 25
Bessel's inequality, 203
Big Bang, 1, 53, 165, 183
Birkhoff, G., 67, 80, 81, 199
birth of time, 76, 79
bit commitment, 169
Bitbol, M., 61, 64, 135, 141
Blokhintsev, D. I., 134
Bogolubov, N. N., 199
Bohm, D., 1, 3, 36, 41, 99, 112–119, 121, 124, 128, 129, 181, 184
Bohr, A., 2, 28, 47–49, 56
Bohr, N., 1, 2, 25, 43, 47, 55, 112, 164, 179
Boole, 67
Boolean algebra, 67, 191
Boolean logic, *see* logic
Boolean mind, 79, 128
Boolean observer, 67, **79**
Borel field, 195
Borel function, *see* function
Borel sets, 194, 195
Born, M., 1, 50
    postulate, 12, 63
    quantum mechanical rule, 60
    rule, 76
Boschi, D., 178–180
bosons, 182

Bourbaki, N., 79
Bouwmeester, D., 180
Branca, S., 180
Branning, D., 178, 179
Brassard, G., 168, 170, 179, 180
Braunstein, S. L., 180
Breidbart, S., 168, 179
Broglie, L. de, 1, 3, 100, 112, 113, 115, 122, 129, 130
    waves, 21
Brown, H. R., 114
Brune, M., 141
Bub, J., 79, 81, 112
bunch, 89
bundle
    Clifford, 121
    cotangent, 129
    spin-Clifford, 121

Caldeira, A. O., 21, 22, 28
canonical mapping, 189
Cartesian product, 188
cascade, *SPS*, 38
Casimir effect, 181, 182
Casimir invariants, 48
Casimir, H. B. G., 181, 184
Cauchy
    sequence in measure, 197
Chau, H. F., 170, 180
Chiao, R. Y., 178
Choquet-Bruhat, Y., 199
classical limit, 14
classicity parameter, 18
Clauser, J. F., 35, 40, 41
collapse
    of wave function, 30
    of wave packet, 29, 50
collective system, 16
Collins, G., 173, 180
commutator product, 213
compatible observables, 8
compatible subspaces, 205
complementarity, 14, 64
    whole-parts, 92
configuration space, 127, 128, 182
conjugate dual, 208
conjugate dual space, 7
conjunction, 72, 152
consciousness, 3, 51, 97
    Boolean, 79
    world — idea, 53
consistency conditions, 152
context
    independence of, 109
context dependent
    HVT of the first kind, 100
contextual dependence, 98, 105

contextual dependence phenomena, 105
coordinate chart, 123
Copenhagen interpretation, 2, *see* interpretation
correlation function, *see* function
Crépeau, C., 170, 180
Cramer, J., 95, 96
cryptography, 159, 168

d'Espagnat, B., 4, 5, 10, 25, 27, 29, 33, 35, 41, 44, 54, 55, 57, 63, 64, 157
    version of Bell's inequalities, 36
Dalibard, I., 41, 180
data, 154
datum, 155
De Leo, S., 7, 216
De Martini, F., 173, 178–180
decoherence, 7, 14, 20, 22
delayed choice experiment, 164
density matrix, 9
Deutsch, D., 27, 168, 179
Dewdney, C., 104, 114
DeWitt, B. S., 25, 28, 54, 59, 60, 64, 65
DeWitt, C., 64
diagram
    Hasse, 84, 88, 89
    sagittal, 84
difference, 188
    symmetric, 188
DiGiuseppe, G., 178, 179
Diosi, L., 146, 149, 155, 157
Dirac current, 117
Dirac measure, 208
Dirac spinor field, 125
disjoint, 188
disjoint subspaces, 205
disjunction, 72
dispersion, 112, 134
dissipation, 16
dissipation effect, 21
Dodson, C. T. J., 216
Doran, C., 126, 130
Dorofeev, V. Yu., 27
Dowker, F., 155, 157
Dowling, J. P., 121, 131
dragging effect, 146
Dreyer, J., 141
Dunne, B. J., 55, 56

Eccles, J. C., 55, 57, 182, 184
Ehrenfest, P.
    theorem, 15, 20
Eibl, M., 180
Einstein, A., 1, 37, 41, 55, 56, 95, 112, 133
Einstein-Podolsky-Rosen, *see EPR*, *see EPR*
Ekert, A. K., 168, 179
Elitzur, A. C., 4, 163, 179
Elitzur-Vaidman proposal, 163
Ellis, J., 146, 149

Emch, G., 26
empty set, 187
empty wave, 119
Enders, A., 178
energy
    non conservation, 117, 145
energy-momentum tensor, 124
Englert, B.-G., 165, 179
ensemble, 32, 102, 112, 133, 134
    homogeneous, 97
environment, 16, 144
environmental approach, 146
EPR, 2, 35, 36, 43, 46, 51, 61, 155, 168, 181–183
equation(s)
    boson, 123
    continuity, 20
    Dirac, 46, 117, 121, 125, 126, 147
        BBT interpretation, 125
    Hamilton-Jacobi, 20, 116, 117, 119, 124
    Klein-Gordon-Fock (*KGFE*), 46, 123, 147
    Maxwell, 46
    n-particle Dirac, 127
    Schrödinger, 29, 30, 44, 46, 50, 60–64, 116–118, 143–145, 151
        master, 60
        nonlinear, 147
        super, 125
    stochastic, 143
    Weyl, 46
equivalence relation, 189
erasers, 165
Esposito, S., 55, 56
EV (Elitzur-Vaidman), 163
Everett, H., 23, 25, 27, 54, 59, 64, 156
evolution
    deterministic, 11
    indeterministic, 11
    Schrödinger, 60, 63

facts, 154
factual, 154
Farris, W., 155–157
Felber, J., 147, 148
Feller, W., 27
fermions, 182
Feynman, R. P., 21, 28, 151, 157
field
    randomly fluctuating, 144
field(s)
    boson, 115
fields, 45
filter, 68
Finkelstein, D., 2, 23, 26, 27, 48, 56, 67, 80, 83, 86, 96
Finkelstein, S. R., 26, 73, 83, 86, 96
first element
    smallest element

    least element, 189
Fock, V. A., 1, 2, 49, 50, 55, 56, 134
Fourier transform, 17
Freidstadt, H., 131
Frenkel, A., 146, 148
frequency
    prediction of, 46
frequency interpretation, *see* interpretation, frequency
FUNC, *see* principle, functional composition
FUNC rule, 109
function
    Borel, 71
    characteristic, 71, 188
    correlation, 39
    frame, 105, 108
    Liouville, 16
    nonexistence of truth value, 107
    Wigner, 17

Gähler, R., 147, 148
Gell-Mann, M., 134, 143, 148, 151–153, 156, 157
George, C., 26, 27
Ghirardi, G. C., 143, 144, 148
Ghirardi-Rimini-Weber localization processes, 183
Gibbs paradox, 181
Gillespie, D. T., 135, 141
Glauber, R. J., 26, 28
Gleason, A. M., 3, 101, 105, 113
GMSZ relations, 159
Golub, R., 147, 148
Graham, N., 64, 65
Graham, R. D., 28
Grangier, P., 41, 113, 180
graph, 83
Grassi, R., 143, 148
Greenberger, D. M., 4, 35, 41, 159, 178
Grib, A. A., 27, 48, 56, 64, 67, 77, 80, 81, 83, 86, 91, 95, 96, 101, 114, 149, 184, 185
Griffiths, R. B., 152, 157
GRW, *see* Ghirardi, Rimini, Weber
GRW effect, 147
Guenin, M., 26
Gueret, Ph., 113, 130, 131
guidance formula, 116
guidance law, 124, 127
Gull, S., 130
Gutler, R., 122

Hörmander, L., 28
Haag, R., 79, 81
Hagley, E., 141
Hamiltonian, 8
    super, 125
Hardy, L., 4, 35, 41, 161, 178–180

Haroche, S., 141
Hartle, J. B., 23, 27, 143, 148, 151–153, 156, 157
Hasse diagram, 74, 77
Hawking, S. W., 64, 146, 149, 157
Heisenberg, W., 1, 2, 44, 49, 55, 56, 112
  uncertainty principle, 118
  uncertainty relation, 134
  uncertainty relations, 83, 89, 90, 178
Heisenberg. W., 49
Heitmann, W., 178
helicity, 61
Hepp, K., 23, 28
Hestenes, D., 117, 121, 122
hidden parameter, 39
hidden variable, 116
hidden variables
  context-dependent, 39
  local, 39
Hilbert space, 7, 16, 18, 24, 30, 72, 74, 76, 79, 83, 102, 104, 107, 110, 129, 135, 154, 155, 198
  rigged, 216
  tensor product, 210
Hilley, B. J., 99, 112, 113, 115–119, 121, 124, 128, 129
histories
  complete family of, 152
  consistent, 152, 153, 155
  inconsistent, 153, 155
histories approach, 144, 151, 183
history, 143, 151, 152
  larger, 152
Holland, P. R., 99, 104, 112–114, 117, 121, 124, 126, 129, 130
holomovement, 119
Holt, R. A., 41
horizontal sum, 89
Horn, M. A., 159
Horne, M. A., 35, 40, 41, 178
Huttner, B., 168, 179
HVT, see theory, hidden variables

improper mixtures, 10
inclusion, 187
indexed family, 188
inequalities, see Bell
  Bell's, 44
  Clauser-Horne, 40
Infeld, L., 146, 149
infimum
  greatest lower bound, 190
internal degree of freedom, 48
interpretation
  absolute Ψ, 54
  Copenhagen (*CI*), 37, 43, 54, 59, 60, 62, 74, 155, 182, 183
  frequency, 7, 11, 23, 50
  histories, 183
  many worlds, 25, 59–62, 64, 183
  objective potentialities existence (*OPE*), 77
  of objectively existing potentialities, 49
  ontological, 115, 128
  quantum logic, 54
  realistic, 112, 128, 183
  statistical — of quantum theory (*SIQT*), 133–138
intersection, 188
meet
  conjunction, 190
Isham, C. J., 157
Israeli, I., 72

Jabs, A., 184
Jahn, R. G., 55, 56
Jammer, M., 4, 5, 67, 79, 80, 112, 141, 143, 148, 199
Jauch, J. M., 26, 67, 68, 79, 80, 101, 109, 114, 199
Jordan, P., 23, 28, 79
Josephson junctions, 1
Jozsa, R., 170, 180
Jung, C. G., 50

Károlyházy, A., 146, 148
Kafatos, M., 178
Kant, I., 3
Kastler, D., 79, 81
Keller, J., 42
Kent, A., 155, 157
Killing vectors, 47
Kochen, S., 3, 101, 105, 109, 111, 113
Kochen-Specker
  paradox, 110
Kolmogorov, 12
Krypianidis, A., 104, 114
Kwiat, P. G., 178

Lagrangian formalism, 126
Landau measurement, *see* measurement, Landau
Landau, L. D., 30, 31, 33
Langlois, D., 170, 180
Laplace determinism, 64
Lasenby, A., 126, 130
lattice
  atomic, 190
  Boolean, 71
  complemented, 191
  complete orthocomplemented, 69
  correspondence with graphs, 83
  definition, 190
  Hasse diagram of, 74
  modular, 191
  non Boolean, 74, 79

non distributive, 55
of propositions, 70, 87
orthocomplemented, 72
quantum, 92
uniquely complemented, 191
Laurikainen, K. V., 57
Leggett, A. J., 21, 22, 28, 149
Leibniz, G. W. von, 77
Leinaas, J., 181, 182, 184
lemma
Gleason's, 105, 107, 108
Lifshitz, E. M., 30, 31, 33
Linden, N., 157
Liouville density, 17
Lo, H-K., 170, 180
locality, 39
logic
Aristotelian, 74
Boolean, 36, 72, 74, 76, 80, 97, 182
Boolean distributive, 76
formal, 67
non Boolean, 74, 76, 182
quantum, 2, 74, 77, 80, 97
macroscopic realizations of, 83
quantum (QL), 67
Logunov, A. A., 199
London, F., 3, 50, 57, 63
Lorentz chart, 126
Lorentz contraction, 45, 55
Lorentz, H. A., 45
boost, 48
invariance, 127, 128
breakdown of, 128
invariant processes, 128
Lounesto, P., 121, 125, 130
lower bound, 190
Lu, J. Y., 42, 129, 130
Ludwig, G., 2, 45, 55
Lukás, B., 146
Lukàcs, B., 148

Maître, X., 141
Maali, A., 141
Mach-Zender interferometer, 164
Mackey, G. W., 199
Mackinnon, L., 129, 130
macro-observable, 16
commuting, 15
macroscopic quantum computer, 101
Maiorino, J. E., 42
Mamayev, S. G., 48, 56, 149
Mandel, L., 178, 179
Mann, A., 180
many pasts, 63
Marlow, A. R., 57
matrix
Dirac, 125
Hermitian, 102
Mattle, K., 180
maximal element, 190
Maximov, Yu. M., 146, 149
Mayers, D., 180
measure
probability, 105
measurement, 8, 14
Hamiltonian, 62
ideal, 8
Landau, 29
non ideal, 9, 30
of the first kind, 8
of the second kind, 9
von Neumann, 8, 62
measures, 194
Mermin, N. D., 108, 114, 161, 179
example, 109
microlocal analysis, 17
Miller Jr., W., 26, 27, 114
minimal element, 190
Minkowski
vacuum, 48
Minkowski inequality, 202
Minkowski spacetime, 47, 48, 79, 123, 128, 165
miracle, 162, 163
mixture, 145
mixture of states, 9
Mjakishev, G. J., 141
modular
lattice of propositions, 87
modular law, 206
Mohanty, S., 146, 149
Mohrhoff, U., 167, 179
momentum
non conservation, 117
Monroe, C., 134, 141
Morris, M. S., 181, 184
Mostepanenko, V. M., 48, 56, 149
Mouken, C. H., 178, 179
Moyal, J. E., 17, 27
MWI (many worlds interpretation), *see* interpretation
Myrheim, R., 181, 182, 184

Nanoupolos, D., 146, 149
Nauenberg, G. M., 28
negative experiments, 163
Nimtz, G., 55, 56, 178
non-polarized particles, 10
nonchaotic system, 18
nonlocal, 119
nonlocal connections, 128
nonlocal influence, 140
nonlocality, 2, 2, 39, 44, 46, 51, 61, 77, 95, 155
active, 128
passive, 128
nonlocality in time, 181

nonseparability, 155
nontrivial topology, 181
normed dual, 208

objective potentialities, 77
observable
    collective, 15, 16
    collective position, 16
    preferable, 22
observable algebra, 8
observer
    Boolean, 91
    consciousness of the, 51
    ultimate, 50
Ocklo, 55
Odzijewicz, A., 146
Omnès, R., 16–18, 21, 27, 143, 144, 148, 151–154, 156, 157
ontological status of physical field, 98
OPE, *see* interpretation
operationalistic philosophy, 45
operator
    adjoint, 211
    algebra, 213
    annihilation, 211
    bounded, 210
    commuting
        complete set of, 109
        maximal complete set of, 109
    continuous, 211
    creation, 211
    projection, 212
    unbounded, 211
or (inclusive), *see* disjunction
order relation, 189
orthocomplement, 92
orthogonal complement, 73
Oziewicz, Z., 42

Palma, G. M., 168, 179
Pan, J. W., 180
Papaliolios, C., 99, 113
paradox, *see* under specific name.
    relativistic, 79
particle(s)
    fermion, 115
partition, 188
parts, 12, 14, 31
Pauli, W., 50, 55, 117
    exclusion principle, 47
    matrices, 102
    principle, 146
Pavsic, M., 112, 114, 121, 130
Pearle, P., 143, 148
Penrose, R., 56, 146, 149, 157, 183, 184
    theory, 48
Peres, A., 42, 54, 55, 168, 179, 180

pertinence, 187
phase space, 15
phenomena, 154
photon, 124
picture
    Heisenberg, 154
    two worlds, 23
Piron, C., 67, 80, 81, 109, 114
Planck, M.
    constant $\hbar$, 83
    constant $\hbar$, 45, 47
    mass, 48
Podolsky, B., 41
Poincaré, H.
    group, 47
Popescu, S., 180
Porteous, I. R., 199
poset, 189, 190
    definition, 189
Post, T., 216
postulate
    statistical, 99
potential
    advanced, 95
    quantum, 119
potential barrier, 126
potentialities, 63
potentiality
    objective existing, 51
pre-Hilbert space, 208
preferable basis, 62, 144
preferred Lorentz frame, 127
Prigogine, I., 26, 27, 143, 144, 147–149
principle
    functional composition (*FUNC*), 109, 110
    of superposition (*PS*), 7
    strong anthropic, 54
probability amplitude, 75
proper mixtures, 10
proper subspace, 205
proposition calculus, 67, 70
pseudo-differential calculus, 17
psychokinetic effect, 128
Putnam, H., 67, 73, 81

QL, *see* logic, quantum
QT, *see* theory, quantum
quantum
    beginning of the universe, 165
    channel, 168
    computers, 3
    cosmology, 59, 181
    cryptography, 168
    ensemble, 134
    indeterminism, 47, 79
    leap, 15
    miracle(s), 159, 160, 162
    object, 45

physics
    fundamentals, 7
    potential, 124
    theory
        orthodox, 137
    theory of motion, 115
    topology, 182
quantum bit commitment
    QBC, 170
quantum cryptography, 1
quantum key distribution, 169
quasi-projectors, 18
quaternionic spaces, 7
quotient set, 189

random classical fields, 124
random field, 147
Rao, K. Ramakrishna, 80, 81
Rapoport, D., 143
Rapoport, D. L., 148
Rauch, H., 104, 114
Raymond, J. P., 141
Recami, E., 130
Redhead, M., 4, 5
reduction, *see* wave packet reduction
    of state vector, 8
Reivelt, K., 41, 96, 131
relation, 189
relative complement, 188
relativity
    special — dogma, 36
relaxation time, 99
Renninger, M., 163, 179
representation
    Heisenberg, 151, 152
    Schrödinger, 151
result(s), 154, 155
Revzen, M., 180
Rimini, A., 143, 144, 148
Rindler vacuum, 48
RIQT (realistic interpretation of quantum theory), *see* interpretation, realistic
Rodrigues Jr., W. A., 7, 42, 55, 56, 112, 114, 121, 122, 125, 129, 130, 216
Roger, G., 41, 180
Roman, P., 27, 199
Rosa, M. A. F., 55, 56, 216
Rosen, N., 41
Rosenfeld, L., 26, 27
Rucker, R., 199

Saari, P., 41, 96, 131
Sachs, R. K., 129, 216
Sallesi, G., 130
Schiller, R., 114
Schilpp, P. A., 56, 164
Schiminovich, S., 26
Schmidt, H., 55, 56, 128, 130

Schrödinger, E., 1, 112
    cat, 144
        cat paradox, 3
    cat paradox, 15, 20, 64, 144, 182
    cat state, 134, 147
    current, 117
    equation, 8, 11, 14
    evolution operator, 151
Schwartz inequality, 202
Scully, M. O., 165, 179
Segal, I. E., 79, 81
Selleri, F., 4, 5, 41, 99, 112–114, 121, 122, 184
sequence, 188
sesquilinear functional, 211
set, 187
Shimony, A., 35, 41, 159
Simmons, G. F., 199
SIQT, *see* interpretation, statistical
Souza, Q. A. G. de, 121, 125, 130, 216
Specker, E. P., 3, 101, 105, 109, 111, 113
spectrum, 152
spin projection, 38, 94
Spliser, D., 26
SQUID, 147
Squires, E. J., 53, 57, 63, 64, 128, 130, 183, 184
standard deviation, 134
Stapp, H. P., 4, 5, 35, 41, 49, 50, 53–57, 128, 130, 151, 157, 183, 184
    formulation of Bell's inequalities, 37, 46
        breakdown of, 39
state
    antisymmetric, 92, 94
    coherent, 19, 20
    mixed, 133
    mixture of, 22, 29, 31, 33
    pure, 14, 133
    stationary, 117
states, 7
statistical operator, 9
Steinberg, A. M., 178
Stern-Gerlach
    apparatus, 104, 160
    experiment, 63
stochasticity, 145
Stroud, C., 28
Stuckelberg, E. C. G., 26
subconsciousness, 51
subensemble, 134
sum of subspaces, 204
super quantum potential, 125
superluminal
    processes, 128
    signal(s), 36, 95, 127
    waves, 95
superselection rule, 79
supremum
    least upper bound, 190

symbol, 17
symmetric wave function, 182
system
  chaotic, 19

tangent bundle, 26
Tarozzi, G., 114
Taylor, A. E., 199
Taylor, M., 27
teleportation, 1, 173
tensor product, 209
theorem
  Gleason's, 105, 109
  Kochen-Specker, 108, 109
  non-go, 101
  von Neumann's, 101, 102, 104
theory
  continuous spontaneous localization, 144
  de Broglie-Bohm, 98
  Dirac, 126
  GRW, 146
  hidden variables, 97–101, 109, 111, 112, 143
    contextuality in, 108
    de Broglie-Bohm, 100, 101, 183
    de Broglie-Bohm non relativistic, 115, 117–120, 126
    de Broglie-Bohm relativistic, 123
    de Broglie-Bohm relativistic many-fermion, 126
    de Broglie-Bohm relativistic *n*-particle, 127
    de Broglie-Bohm relativistic *n*-particle, 128
    of the first kind, 99
    of the second kind, 101
    of the zeroth kind, 101
    realistic, 100
  realistic, 183
Thorne, K. S., 181, 184
time machine, 181
Tipler, J., 57
Tittel, W., 173, 180
Todorov, I. T., 199
Torgerson, J. R., 178, 179
tunneling effect, 178
tunneling time, 126
Tuppinger, D., 104, 114

Ulfbeck, O., 25, 28, 47, 56
ultimate Ego, 182
ultimate Observer, 183
Ünal, N., 121, 131
undistorted progressive waves (*UPW*), 129
union, 188
  join
    disjunction, 190
unitary

  evolution, 30
Unruh, W. G., 56
  radiation, 48
upper bound, 190

vacuum fluctuations, 118
Vaidman, L., 4, 163, 179
van der Merwe, A., 114
Varadarajan, V. S., 81, 199
variable
  hidden, 98, 134
variable(s)
  collective, 144
Vaz Jr., J., 42, 112, 114, 121, 125, 130
vectors, 7
Vernon Jr., F. L., 21, 28, 151, 157
Vigier, J. P., 113, 129–131
von Mises, 12
von Neumann, J., 1, 3, 11, 23, 27–29, 33, 49, 50, 53, 56, 63, 67, 79–81, 101, 112, 113, 182
  axiom, 103, 104
  measurement, *see* measurement, von Neumann
von Weizsäcker, C. F., 43, 55, 164

Walther, H., 165, 179
wave function
  Dirac, 126
  Klein-Gordon-Fock, 123
  localized, 15
  quasi-classical, 20
wave function collapse, 8
wave guides, 178
wave packet collapse, 147
wave packet reduction, 97
wave(s)
  quantum, 115
wavelength
  Compton, 15
  de Broglie, 15
weak modularity postulate
  WMP, 70
Weber, T., 143, 144, 148
Weinfurter, U., 180
Wesley, J. P., 41, 42, 112, 114
Wheeler, J. A., 25, 53, 54, 57, 59, 64, 164, 165, 178, 179
whole, 12, 14, 31
whole, the — and its parts, 7
wholeness, 14, 97
Wiesner, S., 168, 178, 179
Wigner
  functions, 17
Wigner, E. P., 3, 16, 23, 25, 27–29, 31, 33, 47, 50, 56, 57, 79, 143, 148
  formula, 18

formula for probabilities, 152
friend, 3
friend paradox, 53
index function, *see* function
Wooters, W., 180
world
    non Boolean, 76
Wu, H., 129, 216
Wunderlich, C., 141

Yao, A., 170, 180

Yeazell, J., 28
yes-no experiment
    YNE, 67

Zapatrin, R. R., 67, 80, 83, 86, 95, 96, 101, 114, 184
Zeilinger, A., 35, 41, 159, 173, 178, 180
Zeno effect, quantum, 1
zitterbewegung, 126
Zurek, W. H., 21, 28, 144, 148, 151, 157